Progress in Inflammation Research

Series Editor

Prof. Michael J. Parnham PhD
Director of Preclinical Discovery
Centre of Excellence in Macrolide Drug Discovery
GlaxoSmithKline Research Centre Zagreb Ltd.
Prilaz baruna Filipovića 29
HR-10000 Zagreb
Croatia

Advisory Board

Forthcoming titles:

Microarrays in Inflammation, A. Bosio, B. Gerstmayer (Editors), 2008
New Therapeutic Targets in Rheumatoid Arthritis, P.-P. Tak (Editor), 2009
Inflammatory Cardiomyopathy (DCM) – Pathogenesis and Therapy, H.-P. Schultheiß,
    M. Noutsias (Editors), 2009
Th 17 Cells: Role in Inflammation and Autoimmune Disease, B. Ryffel, F. Di Padova
    (Editors), 2009
Occupational Asthma, T. Sigsgaard, D. Heederick (Editors), 2009
Nuclear Receptors and Inflammation, G.Z. Feuerstein, L.P. Freedman, C.K. Glass (Editors),
    2009

(Already published titles see last page.)

Edinburgh

WID

Matrix Metalloproteinases in Tissue Remodelling and Inflammation

Vincent Lagente
Elisabeth Boichot

Editors

Birkhäuser
Basel · Boston · Berlin

Editors

Vincent Lagente
Elisabeth Boichot
INSERM U620
Faculté des Sciences Pharmaceutiques et Biologique
Université de Rennes 1
Avenue du Professeur Léon Bernard 2
35043 Rennes CX
France

Library of Congress Control Number: 2008932707

Bibliographic information published by Die Deutsche Bibliothek
Die Deutsche Bibliothek lists this publication in the Deutsche Nationalbibliografie;
detailed bibliographic data is available in the internet at http://dnb.ddb.de

ISBN 978-3-7643-8584-2 Birkhäuser Verlag AG, Basel – Boston – Berlin

© 2008 Birkhäuser Verlag AG
Basel · Boston · Berlin
P.O. Box 133, CH-4010 Basel, Switzerland
Part of Springer Science+Business Media
Printed on acid-free paper produced from chlorine-free pulp. TCF ∞
Cover design: Markus Etterich, Basel
Cover illustration: see page 47. With friendly permission of Annie Pardo.
Printed in Germany
ISBN 978-3-7643-8584-2   e-ISBN 978-3-7643-8585-9

9 8 7 6 5 4 3 2 1   www.birkhauser.ch

Contents

List of contributors

Chantal Belleguic, Respiratory Medicine Department, Pontchaillou Hospital, 35000 Rennes, France

Maria G. Belvisi, Respiratory Pharmacology Group, Airways Disease Section, Imperial College London, Faculty of Medicine, National Heart and Lung Institute, Dovehouse Street, London SW3 6LY, UK; e-mail: m.belvisi@imperial.ac.uk

Mark A. Birrell, Respiratory Pharmacology Group, Airways Disease Section, Imperial College London, Faculty of Medicine, National Heart and Lung Institute, Dovehouse Street, London SW3 6LY, UK

Elisabeth Boichot, INSERM U620, Faculté de Pharmacie, Université de Rennes 1, 2 avenue du Professeur Léon Bernard, 35000 Rennes, France

Graziella Brinchault, Respiratory Medicine Department, Pontchaillou Hospital, 35000 Rennes, France

Sylvie Caulet-Maugendre, Pathology Department, Pontchaillou Hospital, 35000 Rennes, France

Bruno Clément, INSERM, U-620, Detoxication and Tissue Repair Unit, University of Rennes I, 2 av. Léon Bernard, 35043 Rennes, France; e-mail: bruno.clement@rennes.inserm.fr

Jean-Christophe Copin, Department of Anesthesiology, Pharmacology and Intensive Care, Geneva University Hospitals and Geneva Neuroscience Center, University of Geneva, 1211 Geneva, Switzerland

Kian Fan Chung, Experimental Medicine/Airway Disease Section, National Heart & Lung Institute, Imperial College London, Dovehouse St, London SW3 6LY, UK; e-mail: f.chung@imperial.ac.uk

Benoît Desrues, Respiratory Medicine Department, Pontchaillou Hospital, 35000 Rennes, France

Marie-Pia d'Ortho, INSERM, Unité U841, IRBM, Département Foie-Coeur-Poumon, équipe 6; and Université Paris 12, Faculté de Médecine, IFR10; and AP-HP, Groupe Henri Mondor – Albert Chennevier, Service de Physiologie – Explorations Fonctionnelles, 94000 Créteil, France; e-mail: maria-pia.d-ortho@inserm.fr

Yvan Gasche, Department of Anesthesiology, Pharmacology and Intensive Care, Geneva University Hospitals and Geneva Neuroscience Center, University of Geneva, 1211 Geneva, Switzerland; e-mail: yvan.gasche@medecine.unige.ch

Jean-Yves Gillon, Merck-Serono International S.A., 9 Chemin des Mines, 1211 Geneva 20, Switzerland

Isabelle Guénon, INSERM U620, Faculté de Pharmacie, Université de Rennes 1, 2 avenue du Professeur Léon Bernard, 35000 Rennes, France

Stéphane Jouneau, Respiratory Medicine Department, Pontchaillou Hospital, 35000 Rennes; and INSERM U620, Faculté de Pharmacie, Université de Rennes 1, 2 avenue du Professeur Léon Bernard, 35000 Rennes, France

Vincent Lagente, INSERM U620, Faculté de Pharmacie, Université de Rennes 1, 2 avenue du Professeur Léon Bernard, 35000 Rennes, France; e-mail: vincent.lagente@univ-rennes1.fr

Sum Yee Leung, Department of Respiratory Medicine, Chang Gung Memorial Hospital, Kaohsiung, Taiwan

Guillaume Léveiller Respiratory Medicine Department, Pontchaillou Hospital, 35000 Rennes, France

Jing Lin, Medicine/Cardiology, University of Texas Health Science Center, 7703 Floyd Curl Drive, Mail Code 7872, San Antonio, TX 78229-3900, USA

Merry L. Lindsey, Medicine/Cardiology, University of Texas Health Science Center, 7703 Floyd Curl Drive, Mail Code 7872, San Antonio, TX 78229-3900, USA; e-mail: lindseym@uthscsa.edu

Valeria Muzio, LCG-RBM, Istituto di Ricerche Biomediche "Antoine Marxer", 1 Via Ribes, 10010 Colleretto Giacosa, Italy

Annie Pardo, Facultad de Ciencias, Universidad Nacional Autónoma de México, Apartado Postal 21-630, Coyoacan, México DF, CP 04000, México; e-mail: apardos@servidor.unam.mx

Moisés Selman, Instituto Nacional de Enfermedades Respiratorias Ismael Cosío Villegas, Tlalpan 4502, CP 14080, México DF, México

Catherine le Quément, INSERM U620, Faculté de Pharmacie, Université de Rennes 1, 2 avenue du Professeur Léon Bernard, 35000 Rennes, France

Sissie Wong, Respiratory Pharmacology Group, Airways Disease Section, Imperial College London, Faculty of Medicine, National Heart and Lung Institute, Dovehouse Street, London SW3 6LY, UK

Preface

Matrix metalloproteinases (MMPs), or matrixins, belong to the metzincin super-family of metalloproteinases. MMPs are proteolytic enzymes believed to be impli-cated in many physiological and pathological processes mainly associated with inflammatory reaction. MMPs are zinc and calcium dependent enzymes being able to degrade virtually all extracellular matrix components. MMP synthesis and func-tions are regulated by transcriptional activation, post-transcriptional processing (release of pro-domain, cell surface shedding), and control of activity by a family of endogenous inhibitors collectively known as tissue inhibitors of metalloproteinases (TIMPs). Upon stimulation, many cell types have been identified as producers of MMPs and TIMPs in a context of inflammatory process, strongly suggesting the involvement of MMPs in numerous inflammatory diseases. This mainly includes respiratory pathologies but also diseases of various organs such as liver, central nervous and cardiovascular systems.

From the previous book edited in the same series *Progress in Inflammation Research* by Kevin M.K. Bottomley, David Bradshaw and John S. Nixon in 1999 entitled: *Metalloproteinase as targets for anti-inflammatory drugs*, a spectacular increase in the studies of the role of metalloproteinases on the inflammatory process has been published.

The aim of this new volume is to provide new advances regarding the involve-ment of MMPs in various diseases associated with the inflammatory process. Moreover, the recent development of selective and non-selective inhibitors of MMPs would also provide new insights in the knowledge of the relationship between acti-vation of inflammatory cells and tissue remodelling and propose new therapeutic possibilities to the treatment of inflammatory disease.

The first five chapters are devoted to the airway diseases including acute lung injury and acute respiratory distress syndrome, asthma, chronic obstructive pulmo-nary disease, pulmonary fibrosis and cystic fibrosis. It is now well established that chronic inflammatory response can become detrimental to the airways, causing degradation of the extracellular matrix in the lung. Matrix metalloproteinases are thought to play a critical role in this degradation, and has been the subject of intense research. Two chapters are devoted to the role of MMP and chronic inflammation in cardiovascular diseases. These chapters define cardiac remodelling and pulmonary arterial hypertension, describe MMP-dependent mechanisms that stimulate the remodelling process, and explore future directions and therapeutic potentials in terms of MMP inhibition.

One chapter is devoted to the involvement of MMPs and inflammatory disorders in the central nervous system. Indeed, metalloproteinases, as key modulators of extracellular matrix homeostasis, play a role in the cascades leading to neuronal cell death and tissue regeneration. Yet they may have a detrimental or beneficial role depending on the type and the stage of brain injury.

The last chapter is devoted to the metalloproteinase and extracellular matrix remodelling in inflamed and fibrotic livers. Extracellular matrix remodelling is a complex mechanism of synthesis and degradation of matrix components. Depicting these mechanisms opens the path to the identification of biomarkers and targeted drugs for the reversion of inflamed/fibrotic scar towards a normal architecture and the restoration of normal liver functions.

We thank Hans Detlef Klüber of Birkhäuser Verlag AG for his patience and expert assistance in the preparation of this volume. We are also profoundly grateful to the authors who have contributed to this volume which we believe will provide an important advance in the field of metalloproteinase, extracellular matrix remodelling and chronic inflammatory process.

May 2008

Elisabeth Boichot
Vincent Lagente

The role of MMPs in the inflammatory process of acute lung injury

Sissie Wong, Mark A. Birrell and Maria G. Belvisi

Respiratory Pharmacology Group, Airways Disease Section, Imperial College London, Faculty of Medicine, National Heart and Lung Institute, Dovehouse Street, London SW3 6LY, UK

Abstract

Acute lung injury/acute respiratory distress syndrome (ALI/ARDS) manifests as non-cardiogenic pulmonary oedema, respiratory distress, hypoxemia and results from various processes that directly or indirectly injure the lung (e.g., during sepsis). The innate inflammatory response of the lung to infection, which triggers a cascade of events, including release of proinflammatory cytokines, influx of polymorphonuclear neutrophils in the lung tissue, and production of proteinases, such as matrix metalloproteinases, is central to host defence processes. However, in disease conditions, this chronic inflammatory response can become detrimental to the airways, causing degradation of the extracellular matrix in the lung. Matrix metalloproteinases are thought to play a critical role in this degradation, and has been the subject of intense research. In an attempt to prevent or control disease progression there has been great interest in the development of MMP inhibitors. However, to date, there are no effective therapeutic agents; hence, researchers continue to further elucidate the complex role of these proteinases.

Introduction

The continually growing family of matrix metalloproteinases (MMPs) has been the subject of intense research and has been widely demonstrated to be important in various fields of medicine. MMPs were primarily thought to solely be involved in homeostasis and turnover of the extracellular matrix (ECM), but there has been increasing evidence suggesting that MMPs act on cytokines, chemokines and protein mediators to regulate various aspects of inflammation and immunity [1, 2]. MMPs have been speculated to play a critical role in various inflammatory diseases, such as acute lung injury (ALI), also known as acute respiratory distress syndrome (ARDS) [3], chronic obstructive pulmonary disease (COPD) [4], asthma [5], rheumatoid arthritis [6] and cancer [7]. The precise role of these MMPs in inflammation/immunity is unknown, as it still remains unclear as to whether they are involved in the promotion or reduction of these responses. For instance, *in vitro* studies have demonstrated MMPs to increase the activity of chemokines such as IL-8 [8], but reduce

Matrix Metalloproteinases in Tissue Remodelling and Inflammation,
edited by Vincent Lagente and Elisabeth Boichot

the activity of others, such as ENA78 [9]. Researchers have also demonstrated MMPs to release immobilised chemokine complexes, such as syndecan-1/IL-8 [10], but can convert others, such as monocyte chemoattractant protein-3 (MCP-3), into chemokine receptor antagonists [11]. This chapter explores the role of MMPs and TIMPs in acute lung injury (ALI).

Acute lung injury

The parenchymal lung tissue is comprised of a variety of proteins that make up the extracellular matrix (ECM), which is a complex network of matrix proteins, including collagens, elastin, and proteoglycans. The ECM provides both structural support and regulatory functions for the tissue. Interestingly, it has been demonstrated that the ECM is a source for several inflammatory cell chemotactic factors. In fact, fragments of elastin [12, 13] and collagen [14] have been shown to be chemotactic and could therefore play a role in the inflammatory response.

ALI is an inflammatory condition of the lungs and the most severe form of ALI is ARDS. This disease has been reported to be a frequent complication that emerges in patients after sepsis [15, 16]. In ALI the pulmonary basement membrane is damaged, as a result of the cascade of events that occur in the inflammatory process, such as release of proinflammatory cytokines, influx of polymorphonuclear (PMN) neutrophils in the lung tissue, and production of proteases, such as MMPs. MMPs are thought to play critical roles in lung organogenesis, and these MMPs together with their endogenous inhibitors, tissue inhibitors of matrix metalloproteinases (TIMPs), are produced in the respiratory tract by various cell types, such as alveolar macrophages, neutrophils, fibroblasts, epithelial cells, smooth muscle cells and mast cells. In pathological conditions, the homeostasis of this proteinase–antiproteinase network is disrupted, due to inappropriate secretion of various MMPs by these structural or inflammatory cells during the acute and chronic phases of these diseases. In the lung, these proteolytic MMPs have various targets, such as ECM molecules, chemokines, growth factors, other proteinases and cell surface proteins, such as adhesion molecules. Hence, it is speculated that the imbalance between MMPs and TIMPs plays a critical role in the pathogenesis of inflammatory airways diseases, such as ALI [17–19]. Chronically, this disease may progress to fibrotic lung injury, also known as pulmonary fibrosis [20]. In clinical studies, researchers demonstrated that patients with ARDS had increased MMP-9 levels and an altered ratio of MMP-9 to TIMP-1 levels in the bronchoalveolar lavage (BAL) fluid, compared to controls [21]. Similarly, other researchers also demonstrated an increase in MMP-9 levels [22, 23] and an increase in MMP-2 levels in the BAL fluid of ARDS patients [22]. In addition, to MMP-2 and MMP-9, other MMPs have also been demonstrated to be involved in acute lung injury, where researchers found increased levels of MMP-1, 3 and 8, as well as MMP-2 and 9 in the BAL fluid of patients with ALI [3].

Current *in vitro* research

To aid researchers in elucidating the role of MMPs/TIMPs in inflammatory airway diseases and to enable the development of disease modifying pharmacological agents various techniques have been employed. These include setting up *in vitro* cell-based assays which involve the use of single cell systems provoked by a disease relevant stimulus to produce an inflammatory profile similar to that observed in human pathological conditions.

Lipopolysaccharide (LPS) is a major proinflammatory glycolipid component of the Gram-negative bacterial cell wall of *Escherichia coli*, and has been demonstrated to induce acute pulmonary inflammation associated with neutrophil infiltration [3]. Hence, LPS has been widely used in various *in vitro* and *in vivo* models of inflammation. Cells such as monocytes, macrophages, neutrophils and endothelial cells recognise and respond to the infection caused by invading pathogens. The innate inflammatory response to infection, which results in the release of a range of inflammatory mediators, including TNF-α, IL-1β and IL-6, is central to host defence processes but, in disease conditions, this chronic inflammatory response can become detrimental to the airways.

With the use of cell-based assays, researchers have also been able to investigate the profile of MMPs and TIMPs under normal and 'diseased' conditions. MMP-1 (interstitial collagenases) is known to be undetectable in normal resting tissues [24]. However, there are reports of basal MMP-1 being produced *in vitro* in macrophages [25, 26]. Upon stimulation with LPS, an increase in MMP-1 mRNA expression has been detected in THP-1 monocytic cells and human lung tissue macrophages (HLTMs) [25]. MMP-2 (gelatinase A) is one of the widely researched MMPs along with MMP-9, due to a technique called zymography being readily available. MMP-2 is thought to be involved in the degradation of the ECM, and in the processing of cell receptors and TIMP release [27]. Interestingly, researchers have demonstrated a significant decrease in MMP-2 mRNA expression in HLTMs after LPS stimulation [25], but found the expression levels to be significantly increased in LPS stimulated THP-1 cells [25]. Other researchers have demonstrated an increase in MMP-2 levels in LPS stimulated human bronchial epithelial cells [28]. MMP-3 (stromelysin-1) levels are increased with differentiation of developing mononuclear phagocytes, as MMP-3 levels were found to be negligible in LPS stimulated monocytes, but present in LPS stimulated alveolar macrophages [29, 30]. MMP-7 (matrilysin) functions in defence and repair, and has also been reported to be involved in inflammation, as MMP-7 has been reported to be involved in the activation of TNF-α [31]. MMP-7 expression could not be measured in LPS stimulated THP-1 cells [25, 32] but MMP-7 is expressed in macrophages, as MMP-7 mRNA levels were detectable in unstimulated and LPS stimulated HLTMs [25], and in macrophages from human atherosclerotic lesions [33].

MMP-8 (neutrophil collagenases) is believed to be involved in tissue remodelling during inflammation [34], and the protein is thought to be stored presynthesised as a latent enzyme (pro-MMP-8) within specific granules of neutrophils [35, 36]. However, it is interesting to find evidence in the literature that MMP-8 mRNA was found to be expressed in differentiated peripheral blood monocyte derived macrophages stimulated with CD40 ligand [37], and in monocytes/macrophages in lung tissue samples from bronchiectasis patients [36]. The latter research group suggested that monocyte/macrophages were induced to express MMP-8, which participates in the opening of the migration pathway for inflammatory cells through interstitial tissue, bronchial basement membrane, and the ECM of the affected lung area. Similarly, others have demonstrated an increase in MMP-8 mRNA levels in THP-1 cells and HLTMs stimulated with LPS [25]. Similarly, MMP-9 is also thought to be stored in its latent form (pro-MMP-9) within granules of neutrophils which are thought not to synthesise MMP-9 *de novo* [38]. Pro-inflammatory mediators, such as LPS, TNF-α and IL-8, induce rapid degranulation of neutrophils in whole blood with secretion of pro-MMP-9 [39]. However, it is also speculated that macrophages synthesise MMP-9 (gelatinase B) *de novo* upon stimulation with pro-inflammatory mediators, such as LPS [40]. Others studies have detected MMP-9 mRNA expression in unstimulated THP-1 cells and HLTMs, which was increased after LPS stimulation [25].

The biological role of MMP-10 (stromelysin-2) has been little studied, but has been suggested to be linked with wound healing, as MMP-10 is able to degrade several protein components such as collagen, gelatin, and elastin [41]. Hence, based on this substrate specificity it could also be speculated that MMP-10 may be involved in the pathogenesis of inflammatory airways diseases. Researchers have demonstrated an increase in MMP-10 mRNA expression in LPS stimulated HLTMs [25]. Interestingly, MMP-11 (stromelysin-3) is thought not to degrade the classical MMP substrates that many members of the MMP family degrades, instead it is thought to catalyse the degradation of the serine protease inhibitor, α_1-AT [42]. This MMP is thought to be involved in regulation of cell survival and/or apoptosis [43], and has also been demonstrated to be over expressed in human cancers [7, 44].

MMP-12 (macrophage metalloelastase) has been shown in *in vivo* studies to aid macrophage migration through basement membranes [45], and is known to have a broad substrate specificity, being most active against elastin [46]. Researchers have demonstrated MMP-12 mRNA levels to be upregulated by LPS in THP-1 cells and in HLTMs [25] and MMP-13 (collagenases 3) mRNA levels to be increased in these cell types after LPS. This MMP is also thought to be involved in the degradation of ECM [47]. MMP-14 (membrane-type-1-MMP) is thought to be associated with multiple physiological processes, such as cell migration, angiogenesis, and wound repair [48–50]. MMP-14 has been shown to be a potent activator of pro-MMP-2 [4], and is known to be anchored to the cell membrane rather than being soluble [51]. Basal MMP-14 mRNA levels have been demonstrated to be increased in THP-1 cells and in HLTMs after LPS [25].

It is believed that matrix remodelling is the result, in part, of a shift in the balance between active MMPs *versus* TIMPs, and hence it is thought that coordinated regulation of these proteases and antiproteases is required to maintain tissue architecture [52]. The TIMP family consists of four distinct members, known as TIMP-1–4. Studies have demonstrated that monocytes secrete large quantities of basal levels of TIMP-1, but were unresponsive to LPS, whereas macrophages secrete lower basal levels of TIMP-1, which were found to be upregulated by LPS [25, 26, 53]. It has also been speculated that TIMP-1 may be involved in modulation of inflammatory responses and may also function to stabilise matrix components deposited in the injured lung [54]. Hence, it could also be postulated that TIMPs may also play an essential role in the pathogenesis of inflammatory airways diseases, since TIMPs have also been shown to possess biologic functions which are independent of MMP inhibitory activity, such as induction of MMP expression *in vitro* [55].

Researchers have demonstrated TIMP-2 to be involved in regulation of pro-MMP-2 activation by MMP-14 on cell surfaces of non-malignant monkey kidney epithelial BS-C-1 cells [56]. There is also evidence in the literature that production of TIMP-2 in human fibroblasts is complexed with large amounts of MMP-2, which also are simultaneously secreted by these cells [57–59]. However, other researchers have found TIMP-2 to be secreted *in vitro* from alveolar macrophages in a free or unbound state [53]. Researchers have shown that protein and steady-state mRNA levels of TIMP-2, produced constitutively by alveolar macrophages, were markedly diminished by exposure to LPS, which simultaneously caused marked stimulation in the cell's production of MMP-1 and TIMP-1 [53]. Similarly, other researchers also demonstrated a reduction in TIMP-2 mRNA levels in HLTMs [25].

TIMP-3 is thought to be the only TIMP that binds firmly to the ECM, and through this interaction, TIMP-3 is localised where it can inhibit sheddases or regulate movement through the basement membrane and stroma [60]. Since TIMP-3 is thought to be able to inhibit MMP-1, 2, 3 and 9 [61], researchers have postulated that these MMPs, in the absence of TIMP-3, may damage the collagen and elastin ultrastructure of the lung, leading to the loss of alveolar structure, and result in air space enlargement [52]. TIMP-4 is also known to be able to inhibit MMP-1, 2, 3, 7 and 9 [62], and similar to TIMP-2, it can bind to pro-MMP-2 [63]. Researchers have demonstrated basal levels of TIMP-1, 2 and 3 were highly expressed in human lung, whereas TIMP-4 levels were not detected [64].

Current *in vivo* research

Animal models provide a biologically relevant multicellular system to enable an understanding of the key events in the pathophysiology of inflammatory airways diseases. Although they do not mimic every facet of human complex diseases, they have had a major impact on the investigation of many medical conditions, acting as a bridge between *in vitro* laboratory studies and clinical trials in humans. *In vivo* studies provide insights into the assessment of drug efficacy and provide a framework for the rational and safe design of extensive long-term clinical studies. Many researchers have used LPS instillation in experimental animals to dissect the role of neutrophilic inflammation in the pathogenesis of inflammatory airways diseases [65–68]. Extensive studies investigating short-term LPS exposure in experimental animals have demonstrated that LPS activates alveolar macrophages, among other cells, to produce cytokines, and a rapid but transient neutrophil infiltration into the lung (interstitium, alveoli, and airway) [67]. With technology constantly advancing, researchers have also been able to further investigate the biology of MMPs/TIMPs using a genetic approach, whereby experimental animals have been developed with either over-expression or targeted deletion of specific genes.

Researchers have demonstrated an increase in MMP-2 in the BAL from various animals after LPS insult [69–71], which corresponds with evidence in the literature that there is increased MMP-2 activity in the BAL of patients with ARDS [72, 73]. Similarly, MMP-3 is thought to be involved in the development of experimental acute lung injury, as MMP-3 deficient mice develop less intense lung injury [74]. However, in an *in vivo* model of LPS induced airway neutrophilia model, MMP-3 mRNA levels were found to be below the reliable detection limit of the assay in the vehicle and LPS treated groups [25]. Conversely, these researchers demonstrated MMP-7 mRNA levels to be increased after LPS challenge. This data is consistent with the suggestion that MMP-7 may play a role in regulating the formation of a chemotactic gradient that controls neutrophil influx and activation since neutrophils from MMP-7-null mice have impaired ability to advance from the interstitium into the luminal compartment compared to wild type mice [75]. In addition, other studies have demonstrated that repair of injured tracheal epithelium was impaired in MMP-7-null mice, and suggested that MMP-7 was upregulated in response to injury, possibly to facilitate epithelial cell migration and control repair of the lung epithelium [76]. Hence, it appears that MMP-7 may play a part, along with other upregulated inflammatory biomarkers and mediators, in orchestrating neutrophils in response to injury. However, it appears that MMP-7 may possess both beneficial and harmful roles since it appears to facilitate epithelial cell migration in the airways, but induction of MMP-7 in type II pneumocytes after injury may contribute to alveolar wall damage [76].

Levels of MMP-8 mRNA have been found to be increased after challenging the rats with aerosolised LPS [25]. It could be postulated that the MMP-8 mRNA increase could be a consequence of the neutrophilia observed in this model. However, there is evidence in the literature that MMP-8 protein is presynthesised as pro-MMP-8 and stored subcellularly in the neutrophils [36]. Hence, it could also be postulated that the source of MMP-8 mRNA detected may not be neutrophils, but from other cell types, as this MMP has also been shown to be expressed by monocytes, macrophages, smooth muscle cells, endothelial cells and epithelial cells [25, 36, 37].

Studies have also shown increased MMP-9 activity in the BAL fluid of short-term LPS exposed experimental animals [25, 66], and increased MMP-9 mRNA levels in the rat lung of rats challenged with aerosolised LPS [25]. Since this *in vivo* model exhibits neutrophilia, it could be speculated that the MMP-9 increase may be as a result of an increase in neutrophils. In addition, there is evidence in the literature that neutrophils migrate through pre-existing holes in the airways, upon stimulation of the innate immune response [77], and other researchers have suggested that MMP-9 plays a role in human neutrophil migration *in vitro* [78]. On the contrary, there have been reports in mice, both *in vitro* and *in vivo*, demonstrating that MMP-9 is not required for neutrophil migration [79]. Similarly, other researchers demonstrated that MMP-9 was predominantly expressed in macrophages in histological lung sections [25]. Hence, it could be speculated that the source of the MMP-9 may not be neutrophils, but other inflammatory cells. It is also interesting to note that the MMP-9 profile is similar to the profile of inflammatory biomarkers reported in this model. For example, maximum levels of IL-1β were measured, 8 h after LPS challenge [65], which parallels the MMP-9 profile demonstrated in the same model investigated by other researchers [25, 66]. Hence, it could be postulated that MMP-9 may play a role in the upregulation of IL-1β and *vice versa* [80, 81], and therefore, further stimulating the inflammatory response.

As mentioned earlier, based on the substrate specificity of MMP-10, it could be speculated that this MMP may participate in the degradation of matrix components [41], and hence may be involved in the pathogenesis of inflammatory airways diseases. Interestingly, a significant decrease in MMP-11 mRNA after LPS challenge has been demonstrated [25]. This difference in the gene expression profile between MMP-11 and other MMP family members could imply that this MMP plays a different role compared with other MMPs.

MMP-12 has been demonstrated to aid macrophage migration through basement membranes in *in vivo* studies [45]. Increased levels of MMP-12 mRNA has been demonstrated in rats after LPS challenge [25]. In addition, researchers demonstrated that mice deficient in MMP-12 had less lung injury after immune complex deposition compared to wild type mice [82]. There is also evidence for the activity of MMPs to be involved in the production of other MMPs, for example, Nenan et al. (2005) have shown that addition of MMP-12 can lead to an increase in gelatinase expression [83]. In other studies, MMP-13 mRNA levels were demonstrated to be below the level of detection in the vehicle and LPS challenged rats, suggesting that this MMP may not have much involvement in this model [25].

There is evidence in the literature that basal MMP-14 is highly expressed throughout development from foetal lung to the adult lung [64], which supports the findings by other researchers, who detected basal MMP-14 mRNA levels in the vehicle groups in an *in vivo* model of LPS induced airway neutrophilia [25].

There has also been considerable interest in the role of TIMPs in *in vivo* models of inflammation. Researchers, using a murine model of bleomycin induced pulmonary fibrosis, have found TIMP-1 mRNA to be increased in response to lung injury and suggested that TIMP-1 may be involved in the modulation of the inflammatory response, and may also function to stabilise matrix components deposited in the injured lung [54]. These researchers also demonstrated that TIMP-1 transcripts were most prominent in mononuclear inflammatory cells within the areas of tissue damage, and that the TIMP-1 mRNA expression was spatially restricted to the areas of lung damage. Other researchers demonstrated a significant increase in TIMP-1 mRNA levels in rats after LPS challenge [25]. In a clinical study, researchers found TIMP-1 to be overexpressed in a high percentage of tumours, and suggested that an overexpression of TIMP-1 may be due to the host's response to increased MMP activity, in an attempt to control this activity and retain ECM integrity [84]. It could be postulated that this may also be the reason for the increased TIMP-1 in the *in vivo* models of inflammation observed by researchers. However, the profile of TIMP-2 has been demonstrated to be different to TIMP-1, as researchers found TIMP-2 mRNA levels to be significantly decreased after LPS [25]. In addition, other researchers have demonstrated that instillation of recombinant TIMP-2 into the airways of rats during injury reduced tissue damage [72]. Similarly, TIMP-3 mRNA levels have also been demonstrated to be significantly decreased in rats after LPS challenge [25]. There is also evidence in the literature that absence of TIMP-3 inhibitory activity results in a gradual degradation of ECM, as TIMP-3 null mutant mice demonstrated a progressive enlargement of alveolar air space area [52]. Since TIMP-3 is thought to be able to inhibit MMP-1, 2, 3 and 9 [61], researchers have postulated that these MMPs, in the absence of TIMP-3, may damage the collagen and elastin ultrastructure of the lung, leading to the loss of alveolar structure, and result in air space enlargement [52]. Researchers have also demonstrated a trend towards a reduction in TIMP-4 mRNA levels in rats after LPS challenge [25].

A summary table of MMPs that are known to be produced from different structural and inflammatory cells, either basally or after various stimulants are shown in Table 1.

Despite the vast amount of evidence in the literature that MMPs appear to play a role in chronic inflammatory diseases, many of these reports only investigate the role of one particular MMP, and appear to define a particular MMP to be involved in only one disease. However, recent data suggests that the profile of MMP mRNA expression in three different *in vivo* models of inflammation – LPS induced airway neutrophilia, elastase induced experimental emphysema, and antigen induced airway inflammation – were similar in these models, and suggested that MMPs may play a similar role in different inflammatory airways diseases [25]. Hence, it could be postulated that it is a complex network of MMPs/TIMPs and mediators, rather than individual MMPs/TIMPs that result in the specific pathology seen in particular inflammatory disease states.

Table 1 - MMPs known to be produced from structural and inflammatory cells basally or after various stimuli

Macrophages	Monocytes	Neutrophils	Eosinophils	Mast cells	Epithelial cells	Smooth muscle cells	Fibroblasts
MMP 1 [25, 26]	MMP 1 [25]	MMP 8 [35, 36]	MMP 9 [85, 86]	MMP 1 [87]	MMP 2 [28, 88]	MMP 2 [89]	MMP 1 [90]
MMP 2 [25]	MMP 2 [25]	MMP 9 [38]		MMP 2 [91]	MMP 7 [76]	MMP 8 [37]	MMP 2 [88]
MMP 3 [29, 30]	MMP 7 [92]			MMP 3 [87]	MMP 8 [36, 37]		MMP 3 [90]
MMP 7 [25, 33]	MMP 8 [25]			MMP 9 [93]	MMP 9 [28]		MMP 9 [90]
MMP 8 [25, 36, 37]	MMP 9 [25, 94, 95]				MMP 14 [96]		MMP 14 [97]
MMP 9 [25, 40]	MMP 11 [25]						TIMP 1 [53, 57, 58]
MMP 10 [25]	MMP 12 [25]						TIMP 2 [53, 57–59]
MMP 12 [25]	MMP 13 [25]						
MMP 13 [25]	MMP 14 [25]						
MMP 14 [25]	TIMP 1 [25, 95]						
TIMP 1 [25, 26, 53]	TIMP 2 [25]						
TIMP 2 [25, 53]	TIMP 3 [25]						
TIMP 3 [25]							
TIMP 4 [25]							

Synthetic MMP inhibitors

The potential role of MMPs in the development of inflammatory airways diseases has made these enzymes major targets for therapeutic intervention, and therefore synthetic small molecular weight MMP inhibitors are highly sought [98]. Until now disease treatment using MMP inhibitors has mainly been related to cancer and arthritis, where MMPs were long recognised to play a potential role [99]. The first synthetic inhibitors were developed in the early 1980s, but to date, many MMP inhibitors are still under development, in spite of extensive efforts by almost all major pharmaceutical companies, indicating that the development of MMP inhibitors is very challenging. Among these synthetic inhibitors, there are broad-spectrum inhibitors and specific inhibitors.

In vitro investigations

Although it is known that TNF-α is converted to an active form by TNF-converting enzyme (TACE) [100], there is evidence in the literature that MMP-1, 9 and 17 can process pro-TNF-α into its active form [101]. Since TNF-α is one of the key cytokines in the development of many inflammatory diseases, inhibition of TNF-α converting enzyme (TACE) may have an impact on inflammatory conditions. MMP-2, 3 and 9 is also thought to process the IL-1β precursor into its biologically active form [81], and there is also evidence in the literature that IL-1β can be degraded by MMPs themselves [102]. However, there are reports demonstrating that MMPs are unlikely to be involved in the processing of the IL-1β precursor into its active form in LPS stimulated HLTMs, as an MMP inhibitor was found to have no impact on LPS induced IL-1β levels [103].

In vitro studies have shown that an MMP inhibitor can reduce trans-basement membrane neutrophil migration [78]. Zhang et al. (2004) have shown a dual TACE/MMP inhibitor to reduce LPS induced TNF-α secretion in THP-1 cells, human primary monocytes and human whole blood. However, researchers have demonstrated that MMPs may not be involved with the production of cytokines, in LPS stimulated THP-1 cells [103]. Hence, it could be speculated that, at least in THP-1 cells, the inhibition observed by Zhang et al. (2004) was due to an effect on TACE activity. Interestingly, researchers demonstrated that an MMP inhibitor, which was shown to block the activity of MMP-9 using a fluorogenic activity assay, was observed to have no effect on MMP-9 levels in LPS stimulated THP-1 cells, when measured by zymography [103]. These researchers suggested that the compound was being 'removed' from the active site on MMP-9 during the zymography process, possibly during the denaturing steps or through some of the washing phases.

In vivo investigations

Researchers who investigated a broad spectrum MMP inhibitor observed a dose dependant decrease in LPS induced MMP-9 levels in rats, as measured by zymography [103]. These researchers suggested that the MMP inhibitor was reducing the amount of MMP-9, possibly through inhibition of the *in vivo* activity of one or more MMPs. For example, MMP-2 is able to activate pro-MMP-9 [104], MMP-7 has been shown to activate pro-MMP-2 and pro-MMP-9 [105], and MMP-12 has been demonstrated to activate MMP-2 and MMP-3 [106, 107]. However, this compound was observed to have no effect on BAL and tissue cytokine levels. The lack of inhibitory effect of the MMP inhibitor on TNF-α may suggest that the LPS driven TNF-α processing in the rat lung is solely controlled by TACE. Even though in some cases no effect of an MMP inhibitor was observed on TNF-α levels, there is contradictory evidence in the literature where researchers have demonstrated a reduction in TNF-α levels in LPS challenged mice after treatment with another MMP inhibitor, batimastat [17]. In addition, other groups have reported a reduction in LPS induced plasma levels of TNF-α after administration of other MMP inhibitors [108, 109]. Interestingly, a broad spectrum MMP inhibitor had no effect on the increase in cellular burden in the airway lumen and tissue or activation status of the cells in rats challenged with LPS [103]. Similarly, other researchers also demonstrated no difference in neutrophil levels from LPS challenged mice treated with a different MMP inhibitor, batimastat [17]. Conversely, there is data published by a group that used a dual TACE and MMP inhibitor [110], and by a group that used dual MMP and NE inhibitors [19, 111], that demonstrated cellular inflammation to be inhibited. With the use of these dual inhibitors, it is difficult to determine the role of just the MMPs in cellular inflammation. Hence, it could be speculated that it could be the activity of TACE/NE, and/or a combination with MMPs that is involved in the cellular response to inflammation in the lung.

Conclusion

It remains uncertain which part of the spectrum of MMP activity needs to be inhibited to impact on inflammatory airways diseases, but as there is a wealth of evidence in the literature suggesting that MMPs play an important role in the pathogenesis of acute lung injury, a broad spectrum MMP inhibitor may have therapeutic potential. However, since MMPs also play a role in 'normal' physiological processes, broad spectrum inhibitors may be associated with adverse events. Conversely, while an inhibitor targeting a specific MMP may have reduced side-effects, these compounds may lack therapeutic efficacy where other proteases may be able to substitute for the inhibited protease. Despite the drawbacks encountered, enthusiasm in the development of MMP inhibitors remains high, and the therapeutic potential is expected to grow when positive data are unveiled.

References

1   Greenlee KJ, Werb Z, Kheradmand F (2007) Matrix metalloproteinases in lung: multiple, multifarious, and multifaceted. *Physiol Rev* 87: 69–98

2   Parks WC, Wilson C, Lopez-Boado YS (2004) Matrix metalloproteinases as modulators of inflammation and innate immunity. *Nat Rev Immunol* 4: 617–629

3   Fligiel SE, Standiford T, Fligiel HM, Tashkin D, Strieter RM, Warner RL, Johnson KJ, Varani J (2006) Matrix metalloproteinases and matrix metalloproteinase inhibitors in acute lung injury. *Hum Pathol* 37: 422–430

4   Ohnishi K, Takagi M, Kurokawa Y, Satomi S, Konttinen YT (1998) Matrix metalloproteinase-mediated extracellular matrix protein degradation in human pulmonary emphysema. *Lab Invest* 78: 1077–1087

5   Maisi P, Prikk K, Sepper R, Pirila E, Salo T, Hietanen J, Sorsa T (2002) Soluble membrane-type 1 matrix metalloproteinase (MT1-MMP) and gelatinase A (MMP-2) in induced sputum and bronchoalveolar lavage fluid of human bronchial asthma and bronchiectasis. *APMIS* 110: 771–782

6   Konttinen YT, Ainola M, Valleala H, Ma J, Ida H, Mandelin J, Kinne RW, Santavirta S, Sorsa T, Lopez-Otin C et al (1999) Analysis of 16 different matrix metalloproteinases (MMP-1 to MMP-20) in the synovial membrane: different profiles in trauma and rheumatoid arthritis. *Ann Rheum Dis* 58: 691–697

7   Urbanski SJ, Edwards DR, Maitland A, Leco KJ, Watson A, Kossakowska AE (1992) Expression of metalloproteinases and their inhibitors in primary pulmonary carcinomas. *Br J Cancer* 66: 1188–1194

8   Van Den Steen PE, Proost P, Wuyts A, Van Damme J, Opdenakker G (2000) Neutrophil gelatinase B potentiates interleukin-8 tenfold by aminoterminal processing, whereas it degrades CTAP-III, PF-4, and GRO-alpha and leaves RANTES and MCP-2 intact. *Blood* 96: 2673–2681

9   Van Den Steen PE, Wuyts A, Husson SJ, Proost P, Van Damme J, Opdenakker G (2003) Gelatinase B/MMP-9 and neutrophil collagenase/MMP-8 process the chemokines human GCP-2/CXCL6, ENA-78/CXCL5 and mouse GCP-2/LIX and modulate their physiological activities. *Eur J Biochem* 270: 3739–3749

10   Li Q, Park PW, Wilson CL, Parks WC (2002) Matrilysin shedding of syndecan-1 regulates chemokine mobilization and transepithelial efflux of neutrophils in acute lung injury. *Cell* 111: 635–646

11   McQuibban GA, Gong JH, Tam EM, McCulloch CA, Clark-Lewis I, Overall CM (2000) Inflammation dampened by gelatinase A cleavage of monocyte chemoattractant protein-3. *Science* 289: 1202–1206

12 Senior RM, Griffin GL, Mecham RP (1980) Chemotactic activity of elastin-derived peptides. *J Clin Invest* 66: 859–862

13 Houghton AM, Grisolano JL, Baumann ML, Kobayashi DK, Hautamaki RD, Nehring LC, Cornelius LA, Shapiro SD (2006) Macrophage elastase (matrix metalloproteinase-12) suppresses growth of lung metastases. *Cancer Res* 66: 6149–6155

14 Weathington NM, van Houwelingen AH, Noerager BD, Jackson PL, Kraneveld AD, Galin FS, Folkerts G, Nijkamp FP, Blalock JE (2006) A novel peptide CXCR ligand derived from extracellular matrix degradation during airway inflammation. *Nat Med* 12: 317–323

15 Repine JE (1992) Scientific perspectives on adult respiratory distress syndrome. *Lancet* 339: 466–469

16 Parsons PE, Worthen GS, Moore EE, Tate RM, Henson PM (1989) The association of circulating endotoxin with the development of the adult respiratory distress syndrome. *Am Rev Respir Dis* 140: 294–301

17 Corbel M, Lanchou J, Germain N, Malledant Y, Boichot E, Lagente V (2001) Modulation of airway remodeling-associated mediators by the antifibrotic compound, pirfenidone, and the matrix metalloproteinase inhibitor, batimastat, during acute lung injury in mice. *Eur J Pharmacol* 426: 113–121

18 Woessner JFJ (1991) Matrix metalloproteinases and their inhibitors in connective tissue remodeling. *FASEB J* 5: 2145–2154

19 Carney DE, McCann UG, Schiller HJ, Gatto LA, Steinberg J, Picone AL, Nieman GF (2001) Metalloproteinase inhibition prevents acute respiratory distress syndrome. *J Surg Res* 99: 245–252

20 Ware LB, Matthay MA (2000) The acute respiratory distress syndrome. *N Engl J Med* 342: 1334–1349

21 Lanchou J, Corbel M, Tanguy M, Germain N, Boichot E, Theret N, Clement B, Lagente V, Malledant Y (2003) Imbalance between matrix metalloproteinases (MMP-9 and MMP-2) and tissue inhibitors of metalloproteinases (TIMP-1 and TIMP-2) in acute respiratory distress syndrome patients. *Crit Care Med* 31: 536–542

22 Torii K, Iida K, Miyazaki Y, Saga S, Kondoh Y, Taniguchi H, Taki F, Takagi K, Matsuyama M, Suzuki R (1997) Higher concentrations of matrix metalloproteinases in bronchoalveolar lavage fluid of patients with adult respiratory distress syndrome. *Am J Respir Crit Care Med* 155: 43–46

23 Ricou B, Nicod L, Lacraz S, Welgus HG, Suter PM, Dayer JM (1996) Matrix metalloproteinases and TIMP in acute respiratory distress syndrome. *Am J Respir Crit Care Med* 154: 346–352

24 Pardo A, Selman M (2005) MMP-1: the elder of the family. *Int J Biochem Cell Biol* 37: 283–288

25 Wong S, Belvisi MG, Birrell MA (2005) Profiling of MMP/TIMP gene expression in human and rodent lung samples, and in models of aiway inflammation. *Proc Am Thor Soc* 2: A73

26 Campbell EJ, Cury JD, Lazarus CJ, Welgus HG (1987) Monocyte procollagenase and

tissue inhibitor of metalloproteinases. Identification, characterization, and regulation of secretion. *J Biol Chem* 262: 15862–15868

27   Inaki N, Tsunezuka Y, Kawakami K, Sato H, Takino T, Oda M, Watanabe G (2004) Increased matrix metalloproteinase-2 and membrane type 1 matrix metalloproteinase activity and expression in heterotopically transplanted murine tracheas. *J Heart Lung Transplant* 23: 218–227

28   Yao PM, Buhler JM, D'Ortho MP, Lebargy F, Delclaux C, Harf A, Lafuma C (1996) Expression of matrix metalloproteinase gelatinases A and B by cultured epithelial cells from human bronchial explants. *J Biol Chem* 271: 15580–15589

29   Campbell EJ, Cury JD, Shapiro SD, Goldberg GI, Welgus HG (1991) Neutral proteinases of human mononuclear phagocytes. Cellular differentiation markedly alters cell phenotype for serine proteinases, metalloproteinases, and tissue inhibitor of metalloproteinases. *J Immunol* 146: 1286–1293

30   Welgus HG, Campbell EJ, Cury JD, Eisen AZ, Senior RM, Wilhelm SM, Goldberg GI (1990) Neutral metalloproteinases produced by human mononuclear phagocytes. Enzyme profile, regulation, and expression during cellular development. *J Clin Invest* 86: 1496–1502

31   Haro H, Crawford HC, Fingleton B, Shinomiya K, Spengler DM, Matrisian LM (2000) Matrix metalloproteinase-7-dependent release of tumor necrosis factor-alpha in a model of herniated disc resorption. *J Clin Invest* 105: 143–150

32   Klein RD, Borchers AH, Sundareshan P, Bougelet C, Berkman MR, Nagle RB, Bowden GT (1997) Interleukin-1beta secreted from monocytic cells induces the expression of matrilysin in the prostatic cell line LNCaP. *J Biol Chem* 272: 14188–14192

33   Halpert I, Sires UI, Roby JD, Potter-Perigo S, Wight TN, Shapiro SD, Welgus HG, Wickline SA, Parks WC (1996) Matrilysin is expressed by lipid-laden macrophages at sites of potential rupture in atherosclerotic lesions and localizes to areas of versican deposition, a proteoglycan substrate for the enzyme. *Proc Natl Acad Sci USA* 93: 9748–9753

34   Owen CA, Hu Z, Lopez-Otin C, Shapiro SD (2004) Membrane-bound matrix metalloproteinase-8 on activated polymorphonuclear cells is a potent, tissue inhibitor of metalloproteinase-resistant collagenase and serpinase. *J Immunol* 172: 7791–7803

35   Hasty KA, Hibbs MS, Kang AH, Mainardi CL (1986) Secreted forms of human neutrophil collagenase. *J Biol Chem* 261: 5645–5650

36   Prikk K, Maisi P, Pirila E, Sepper R, Salo T, Wahlgren J, Sorsa T (2001) *In vivo* collagenase-2 (MMP-8) expression by human bronchial epithelial cells and monocytes/macrophages in bronchiectasis. *J Pathol* 194: 232–238

37   Herman MP, Sukhova GK, Libby P, Gerdes N, Tang N, Horton DB, Kilbride M, Breitbart RE, Chun M, Schonbeck U (2001) Expression of neutrophil collagenase (matrix metalloproteinase-8) in human atheroma: a novel collagenolytic pathway suggested by transcriptional profiling. *Circulation* 104: 1899–1904

38   Owen CA, Hu Z, Barrick B, Shapiro SD (2003) Inducible expression of tissue inhibitor of metalloproteinases-resistant matrix metalloproteinase-9 on the cell surface of neutrophils. *Am J Respir Cell Mol Biol* 29: 283–294

39 Pugin J, Widmer MC, Kossodo S, Liang CM, Preas HL, Suffredini AF (1999) Human neutrophils secrete gelatinase B *in vitro* and *in vivo* in response to endotoxin and proinflammatory mediators. *Am J Respir Cell Mol Biol* 20: 458–464

40 Gibbs DF, Warner RL, Weiss SJ, Johnson KJ, Varani J (1999) Characterization of matrix metalloproteinases produced by rat alveolar macrophages. *Am J Respir Cell Mol Biol* 20: 1136–1144

41 Rechardt O, Elomaa O, Vaalamo M, Paakkonen K, Jahkola T, Hook-Nikanne J, Hembry RM, Hakkinen L, Kere J, Saarialho-Kere U (2000) Stromelysin-2 is upregulated during normal wound repair and is induced by cytokines. *J Invest Dermatol* 115: 778–787

42 Deng H, Guo RF, Li WM, Zhao M, Lu YY (2005) Matrix metalloproteinase 11 depletion inhibits cell proliferation in gastric cancer cells. *Biochem Biophys Res Commun* 326: 274–281

43 Wu E, Mari BP, Wang F, Anderson IC, Sunday ME, Shipp MA (2001) Stromelysin-3 suppresses tumor cell apoptosis in a murine model. *J Cell Biochem* 82: 549–555

44 Basset P, Bellocq JP, Wolf C, Stoll I, Hutin P, Limacher JM, Podhajcer OL, Chenard MP, Rio MC, Chambon P (1990) A novel metalloproteinase gene specifically expressed in stromal cells of breast carcinomas. *Nature* 348: 699–704

45 Shipley JM, Wesselschmidt RL, Kobayashi DK, Ley TJ, Shapiro SD (1996) Metalloelastase is required for macrophage-mediated proteolysis and matrix invasion in mice. *Proc Natl Acad Sci USA* 93: 3942–3946

46 Saarialho-Kere U, Kerkela E, Jeskanen L, Hasan T, Pierce R, Starcher B, Raudasoja R, Ranki A, Oikarinen A, Vaalamo M (1999) Accumulation of matrilysin (MMP-7) and macrophage metalloelastase (MMP-12) in actinic damage. *J Invest Dermatol* 113: 664–672

47 Freije JM, Diez-Itza I, Balbin M, Sanchez LM, Blasco R, Tolivia J, Lopez-Otin C (1994) Molecular cloning and expression of collagenase-3, a novel human matrix metalloproteinase produced by breast carcinomas. *J Biol Chem* 269: 16766–16773

48 Bauvois B (2004) Transmembrane proteases in cell growth and invasion: new contributors to angiogenesis? *Oncogene* 23: 317–329

49 Deryugina EI, Ratnikov BI, Strongin AY (2003) Prinomastat, a hydroxamate inhibitor of matrix metalloproteinases, has a complex effect on migration of breast carcinoma cells. *Int J Cancer* 104: 533–541

50 Hotary KB, Allen ED, Brooks PC, Datta NS, Long MW, Weiss SJ (2003) Membrane type I matrix metalloproteinase usurps tumor growth control imposed by the three-dimensional extracellular matrix. *Cell* 114: 33–45

51 Genis L, Galvez BG, Gonzalo P, Arroyo AG (2006) MT1-MMP: universal or particular player in angiogenesis? *Cancer Metastasis Rev* 25: 77–86

52 Leco KJ, Waterhouse P, Sanchez OH, Gowing KL, Poole AR, Wakeham A, Mak TW, Khokha R (2001) Spontaneous air space enlargement in the lungs of mice lacking tissue inhibitor of metalloproteinases-3 (TIMP-3). *J Clin Invest* 108: 817–829

53 Shapiro SD, Kobayashi DK, Welgus HG (1992) Identification of TIMP-2 in human

alveolar macrophages. Regulation of biosynthesis is opposite to that of metalloproteinases and TIMP-1. *J Biol Chem* 267: 13890–13894

54   Madtes DK, Elston AL, Kaback LA, Clark JG (2001) Selective induction of tissue inhibitor of metalloproteinase-1 in bleomycin-induced pulmonary fibrosis. *Am J Respir Cell Mol Biol* 24: 599–607

55   Clark IM, Powell LK, Cawston TE (1994) Tissue inhibitor of metalloproteinases (TIMP-1) stimulates the secretion of collagenase from human skin fibroblasts. *Biochem Biophys Res Commun* 203: 874–880

56   Hernandez-Barrantes S, Toth M, Bernardo MM, Yurkova M, Gervasi DC, Raz Y, Sang QA, Fridman R (2000) Binding of active (57 kDa) membrane type 1-matrix metalloproteinase (MT1-MMP) to tissue inhibitor of metalloproteinase (TIMP)-2 regulates MT1-MMP processing and pro-MMP-2 activation. *J Biol Chem* 275: 12080–12089

57   Howard EW, Bullen EC, Banda MJ (1991) Preferential inhibition of 72- and 92-kDa gelatinases by tissue inhibitor of metalloproteinases-2. *J Biol Chem* 266: 13070–13075

58   Howard EW, Bullen EC, Banda MJ (1991) Regulation of the autoactivation of human 72-kDa progelatinase by tissue inhibitor of metalloproteinases-2. *J Biol Chem* 266: 13064–13069

59   Stetler-Stevenson WG, Brown PD, Onisto M, Levy AT, Liotta LA (1990) Tissue inhibitor of metalloproteinases-2 (TIMP-2) mRNA expression in tumor cell lines and human tumor tissues. *J Biol Chem* 265: 13933–13938

60   Woessner JF Jr (2001) That impish TIMP: the tissue inhibitor of metalloproteinases-3. *J Clin Invest* 108: 799–800

61   Apte SS, Olsen BR, Murphy G (1996) The gene structure of tissue inhibitor of metalloproteinases (TIMP)-3 and its inhibitory activities define the distinct TIMP gene family. *J Biol Chem* 271: 2874

62   Liu YE, Wang M, Greene J, Su J, Ullrich S, Li H, Sheng S, Alexander P, Sang QA, Shi YE (1997) Preparation and characterization of recombinant tissue inhibitor of metalloproteinase 4 (TIMP-4). *J Biol Chem* 272: 20479–20483

63   Bigg HF, Shi YE, Liu YE, Steffensen B, Overall CM (1997) Specific, high affinity binding of tissue inhibitor of metalloproteinases-4 (TIMP-4) to the COOH-terminal hemopexin-like domain of human gelatinase A. TIMP-4 binds progelatinase A and the COOH-terminal domain in a similar manner to TIMP-2. *J Biol Chem* 272: 15496–15500

64   Ryu J, Vicencio AG, Yeager ME, Kashgarian M, Haddad GG, Eickelberg O (2005) Differential expression of matrix metalloproteinases and their inhibitors in human and mouse lung development. *Thromb Haemost* 94: 175–183

65   Haddad EB, Birrell M, McCluskie K, Ling A, Webber SE, Foster ML, Belvisi MG (2001) Role of p38 MAP kinase in LPS-induced airway inflammation in the rat. *Br J Pharmacol* 132: 1715–1724

66   Underwood DC, Osborn RR, Bochnowicz S, Webb EF, Rieman DJ, Lee JC, Romanic AM, Adams JL, Hay DW, Griswold DE (2000) SB 239063, a p38 MAPK inhibitor, reduces neutrophilia, inflammatory cytokines, MMP-9, and fibrosis in lung. *Am J Physiol Lung Cell Mol Physiol* 279: L895–L902

67 Vernooy JH, Dentener MA, van Suylen RJ, Buurman WA, Wouters EF (2001) Intratracheal instillation of lipopolysaccharide in mice induces apoptosis in bronchial epithelial cells: no role for tumor necrosis factor-alpha and infiltrating neutrophils. *Am J Respir Cell Mol Biol* 24: 569–576

68 van Helden HP, Kuijpers WC, Steenvoorden D, Go C, Bruijnzeel PL, van EM, Haagsman HP (1997) Intratracheal aerosolization of endotoxin (LPS) in the rat: a comprehensive animal model to study adult (acute) respiratory distress syndrome. *Exp Lung Res* 23: 297–316

69 D'Ortho MP, Jarreau PH, Delacourt C, quin-Mavier I, Levame M, Pezet S, Harf A, Lafuma C (1994) Matrix metalloproteinase and elastase activities in LPS-induced acute lung injury in guinea pigs. *Am J Physiol* 266: L209–L216

70 Corbel M, Lagente V, Theret N, Germain N, Clement B, Boichot E (1999) Comparative effects of betamethasone, cyclosporin and nedocromil sodium in acute pulmonary inflammation and metalloproteinase activities in bronchoalveolar lavage fluid from mice exposed to lipopolysaccharide. *Pulm Pharmacol Ther* 12: 165–171

71 Coimbra R, Melbostad H, Loomis W, Porcides RD, Wolf P, Tobar M, Hoyt DB (2006) LPS-induced acute lung injury is attenuated by phosphodiesterase inhibition: effects on proinflammatory mediators, metalloproteinases, NF-kappaB, and ICAM-1 expression. *J Trauma* 60: 115–125

72 Gibbs DF, Shanley TP, Warner RL, Murphy HS, Varani J, Johnson KJ (1999) Role of matrix metalloproteinases in models of macrophage-dependent acute lung injury. Evidence for alveolar macrophage as source of proteinases. *Am J Respir Cell Mol Biol* 20: 1145–1154

73 Windsor AC, Mullen PG, Fowler AA, Sugerman HJ (1993) Role of the neutrophil in adult respiratory distress syndrome. *Br J Surg* 80: 10–17

74 Warner RL, Beltran L, Younkin EM, Lewis CS, Weiss SJ, Varani J, Johnson KJ (2001) Role of stromelysin 1 and gelatinase b in experimental acute lung injury. *Am J Respir Cell Mol Biol* 24: 537–544

75 Parks WC (2003) Matrix metalloproteinases in lung repair. *Eur Respir J* (Suppl) 44: 36s–38s

76 Dunsmore SE, Saarialho-Kere UK, Roby JD, Wilson CL, Matrisian LM, Welgus HG, Parks W (1998) Matrilysin expression and function in airway epithelium. *J Clin Invest* 102: 1321–1331

77 Walker DC, Behzad AR, Chu F (1995) Neutrophil migration through preexisting holes in the basal laminae of alveolar capillaries and epithelium during streptococcal pneumonia. *Microvasc Res* 50: 397–416

78 Delclaux C, Delacourt C, D'Ortho MP, Boyer V, Lafuma C, Harf A (1996) Role of gelatinase B and elastase in human polymorphonuclear neutrophil migration across basement membrane. *Am J Respir Cell Mol Biol* 14: 288–295

79 Betsuyaku T, Shipley JM, Liu Z, Senior RM (1999) Neutrophil emigration in the lungs, peritoneum, and skin does not require gelatinase B. *Am J Respir Cell Mol Biol* 20: 1303–1309

80   Lappalainen U, Whitsett JA, Wert SE, Tichelaar JW, Bry K (2005) Interleukin-1beta causes pulmonary inflammation, emphysema, and airway remodeling in the adult murine lung. *Am J Respir Cell Mol Biol* 32: 311–318

81   Schonbeck U, Mach F, Libby P (1998) Generation of biologically active IL-1 beta by matrix metalloproteinases: a novel caspase-1-independent pathway of IL-1 beta processing. *J Immunol* 161: 3340–3346

82   Warner RL, Lewis CS, Beltran L, Younkin EM, Varani J, Johnson KJ (2001) The role of metalloelastase in immune complex-induced acute lung injury. *Am J Pathol* 158: 2139–2144

83   Nenan S, Boichot E, Lagente V, Bertrand CP (2005) Macrophage elastase (MMP-12): a pro-inflammatory mediator? *Mem Inst Oswaldo Cruz* 100 (Suppl 1): 167–172

84   Heppner KJ, Matrisian LM, Jensen RA, Rodgers WH (1996) Expression of most matrix metalloproteinase family members in breast cancer represents a tumor-induced host response. *Am J Pathol* 149: 273–282

85   Ohno I, Ohtani H, Nitta Y, Suzuki J, Hoshi H, Honma M, Isoyama S, Tanno Y, Tamura G, Yamauchi K et al (1997) Eosinophils as a source of matrix metalloproteinase-9 in asthmatic airway inflammation. *Am J Respir Cell Mol Biol* 16: 212–219

86   Schwingshackl A, Duszyk M, Brown N, Moqbel R (1999) Human eosinophils release matrix metalloproteinase-9 on stimulation with TNF-alpha. *J Allergy Clin Immunol* 104: 983–989

87   Jeong WI, Do SH, Jeong DH, Hong IH, Park JK, Ran KM, Yang HJ, Yuan DW, Kim SB, Cha MS et al (2006) Kinetics of MMP-1 and MMP-3 produced by mast cells and macrophages in liver fibrogenesis of rat. *Anticancer Res* 26: 3517–3526

88   Senior RM, Griffin GL, Fliszar CJ, Shapiro SD, Goldberg GI, Welgus HG (1991) Human 92- and 72-kilodalton type IV collagenases are elastases. *J Biol Chem* 266: 7870–7875

89   Huhtala P, Chow LT, Tryggvason K (1990) Structure of the human type IV collagenase gene. *J Biol Chem* 265: 11077–11082

90   Fisher GJ, Datta SC, Talwar HS, Wang ZQ, Varani J, Kang S, Voorhees JJ (1996) Molecular basis of sun-induced premature skin ageing and retinoid antagonism. *Nature* 379: 335–339

91   Kanbe N, Tanaka A, Kanbe M, Itakura A, Kurosawa M, Matsuda H (1999) Human mast cells produce matrix metalloproteinase 9. *Eur J Immunol* 29: 2645–2649

92   Busiek DF, Ross FP, McDonnell S, Murphy G, Matrisian LM, Welgus HG (1992) The matrix metalloprotease matrilysin (PUMP) is expressed in developing human mononuclear phagocytes. *J Biol Chem* 267: 9087–9092

93   Fang KC, Wolters PJ, Steinhoff M, Bidgol A, Blount JL, Caughey GH (1999) Mast cell expression of gelatinases A and B is regulated by kit ligand and TGF-beta. *J Immunol* 162: 5528–5535

94   Saja K, Chatterjee U, Chatterjee BP, Sudhakaran PR (2007) Activation dependent expression of MMPs in peripheral blood mononuclear cells involves protein kinase A. *Mol Cell Biochem* 296: 185–192

95   Johnatty RN, Taub DD, Reeder SP, Turcovski-Corrales SM, Cottam DW, Stephenson

TJ, Rees RC (1997) Cytokine and chemokine regulation of proMMP-9 and TIMP-1 production by human peripheral blood lymphocytes. *J Immunol* 158: 2327–2333

96 Kadono Y, Okada Y, Namiki M, Seiki M, Sato H (1998) Transformation of epithelial Madin-Darby canine kidney cells with p60(v-src) induces expression of membrane-type 1 matrix metalloproteinase and invasiveness. *Cancer Res* 58: 2240–2244

97 Seltzer JL, Lee AY, Akers KT, Sudbeck B, Southon EA, Wayner EA, Eisen AZ (1994) Activation of 72-kDa type IV collagenase/gelatinase by normal fibroblasts in collagen lattices is mediated by integrin receptors but is not related to lattice contraction. *Exp Cell Res* 213: 365–374

98 Skiles JW, Gonnella NC, Jeng AY (2001) The design, structure, and therapeutic application of matrix metalloproteinase inhibitors. *Curr Med Chem* 8: 425–474

99 Daheshia M (2005) Therapeutic inhibition of matrix metalloproteinases for the treatment of chronic obstructive pulmonary disease (COPD). *Curr Med Res Opin* 21: 587–594

100 Black RA, Rauch CT, Kozlosky CJ, Peschon JJ, Slack JL, Wolfson MF, Castner BJ, Stocking KL, Reddy P, Srinivasan S et al (1997) A metalloproteinase disintegrin that releases tumour-necrosis factor-alpha from cells. *Nature* 385: 729–733

101 English WR, Puente XS, Freije JM, Knauper V, Amour A, Merryweather A, Lopez-Otin C, Murphy G (2000) Membrane type 4 matrix metalloproteinase (MMP17) has tumor necrosis factor-alpha convertase activity but does not activate pro-MMP2. *J Biol Chem* 275: 14046–14055

102 Ito A, Mukaiyama A, Itoh Y, Nagase H, Thogersen IB, Enghild JJ, Sasaguri Y, Mori Y (1996) Degradation of interleukin 1beta by matrix metalloproteinases. *J Biol Chem* 271: 14657–14660

103 Birrell MA, Wong S, Dekkak A, De AJ, Haj-Yahia S, Belvisi MG.(2006) Role of matrix metalloproteinases in the inflammatory response in human airway cell-based assays and in rodent models of airway disease. *J Pharmacol Exp Ther* 318: 741–750

104 Fridman R, Toth M, Pena D, Mobashery S (1995) Activation of progelatinase B (MMP-9) by gelatinase A (MMP-2). *Cancer Res* 55: 2548–2555

105 Wang FQ, So J, Reierstad S, Fishman DA (2005) Matrilysin (MMP-7) promotes invasion of ovarian cancer cells by activation of progelatinase. *Int J Cancer* 114: 19–31

106 Shapiro SD, Griffin GL, Gilbert DJ, Jenkins NA, Copeland NG, Welgus HG, Senior RM, Ley TJ (1992) Molecular cloning, chromosomal localization, and bacterial expression of a murine macrophage metalloelastase. *J Biol Chem* 267: 4664–4671

107 Shapiro SD, Kobayashi DK, Ley TJ (1993) Cloning and characterization of a unique elastolytic metalloproteinase produced by human alveolar macrophages. *J Biol Chem* 268: 23824–23829

108 Morimoto Y, Nishikawa K, Ohashi M (1997) KB-R7785, a novel matrix metalloproteinase inhibitor, exerts its antidiabetic effect by inhibiting tumor necrosis factor-alpha production. *Life Sci* 61: 795–803

109 Solorzano CC, Ksontini R, Pruitt JH, Auffenberg T, Tannahill C, Galardy RE, Schultz GP, MacKay SL, Copeland EM III, Moldawer LL (1997) A matrix metalloproteinase

inhibitor prevents processing of tumor necrosis factor alpha (TNF alpha) and abrogates endotoxin-induced lethality. *Shock* 7: 427–431

110 Trifilieff A, Walker C, Keller T, Kottirsch G, Neumann U (2002) Pharmacological profile of PKF242-484 and PKF241-466, novel dual inhibitors of TNF-alpha converting enzyme and matrix metalloproteinases, in models of airway inflammation. *Br J Pharmacol* 135: 1655–1664

111 McCann U G, Gatto LA, Searles B, Carney DE, Lutz CJ, Picone AL, Schiller HJ, Nieman GF (1999) Matrix metalloproteinase inhibitor: differential effects on pulmonary neutrophil and monocyte sequestration following cardiopulmonary bypass. *J Extra Corpor Technol* 31: 67–75

Matrix metalloproteinases in airways inflammation of asthma and chronic obstructive pulmonary disease

Sum-Yee Leung and Kian Fan Chung

Experimental Medicine/Airway Disease Section, National Heart & Lung Institute, Imperial College London, Dovehouse St, London SW3 6LY, UK

Abstract

Matrix metalloproteinases (MMPs) are produced in the respiratory tract from different inflammatory and structural cells and are involved in wound healing, inflammatory cell trafficking and tissue remodelling and repair. MMPs are tightly regulated and their effects are counterbalanced by their physiological inhibitors, tissue inhibitors of MMPs (TIMPs). Increased secretion and expression of MMPs have been reported in asthma and chronic obstructive pulmonary disease (COPD), which are chronic inflammatory lung diseases that lead to chronic airflow obstruction associated with significant mortality and morbidity. Cytokines and growth factors which are involved in these inflammatory processes interact directly with MMPs, leading to a regulation of their expression or changes in their biological activities through enzymatic cleavage. Different MMPs play a specific role with variations in different lung diseases. In COPD, MMP 1, MMP 9 and MMP 12 from lung macrophages play an important role in degradation of matrix leading to emphysema. MMPs may represent relevant therapeutic targets for many diseases. However, their contribution is potentially complex since MMPs may have both beneficial as well as deleterious effects. Thus, some actions may lead to pro-inflammatory effects, while others may also cause anti-inflammatory effects. Therefore, the precise role of these MMPs in airways disease with airflow obstruction need to be clarified carefully before selective MMP inhibitors are tried for therapeutic aims in these diseases.

Introduction

MMPs are a family of zinc-dependent proteases that were initially identified by their ability to degrade collagen [1]. MMPs can be broadly classified on the basis of substrate specificity into collagenases (MMP-1, 8, 13 and 18), gelatinases (MMP-2 and 9), stromelysins (MMP-3, 10, 11), elastases (MMP-7 and 12), and membrane type MMPs (MT-MMPs, MMP-14, 15, 16 and 17) which are surface-anchored. MMP-1 (interstitial collagenase), MMP-8 (neutrophil collagenase), MMP-13 (collagenase 3), and MMP-14 (MT1-MMP) can cleave the triple helix of native type I collagen [2], the primary architectural collagen of the lung [3].

Since MMPs can cause significant damage, they are tightly regulated. Firstly, they are rarely stored but require gene transcription before secretion, the exception being neutrophil MMP-8 and 9. Secondly, they are either secreted as pro-enzymes that require proteolytic cleavage or, in the case of MT-MMPs, activated intracellularly by pro-protein convertases such as furin, that exposes the catalytic cleft, a mechanism known as the cysteine switch [4]. Thirdly, specific inhibitors of MMPs – the tissue inhibitors of metalloproteinases (TIMPs) – are secreted and bind MMPs to prevent enzymatic activity [5]. The balance of MMPs to TIMPs therefore determines matrix turnover, where either an excess of MMPs or a deficit of TIMPs may result in excess ECM degradation. Finally, MMPs can be compartmentalised in close proximity to the cell.

In this chapter, we will review the expression of MMPs in the lung and airways in relation to asthma and COPD, and discuss the potential role of MMPs in the pathophysiology of these airway diseases, particularly in relation to the inflammatory and remodelling processes.

Asthma: definitions and pathophysiology

Asthma is defined as an inflammatory disease of the airways that is associated with an abnormal response mechanisms of the airway smooth muscle that leads to episodes of airway narrowing, making asthma a disease where the airways are prone to narrow excessively in response to many provoking inhaled stimuli. The chronic airway inflammation of asthma is characterised by an infiltration of CD4+ T cells, eosinophils, macrophages/monocytes, and mast cells, and sometimes neutrophils, possibly a marker of severity. A specific class of T cells referred to as T-helper type (Th2) cells orchestrate and perpetuate inflammation and remodelling through the secretion of many cytokines including IL-4, IL-5, IL-9 and IL-13. Each of these play distinct roles with IL-4 regulating IgE production by B cells, while IL-13 is involved in bronchial hyperresponsiveness, subepithelial fibrosis and eosinophilic inflammation. Acute-on-chronic inflammation may be observed with acute exacerbations of the disease, with an increase in eosinophils and neutrophils in the airway submucosa and release of mediators such as histamine and cysteinyl leukotrienes from eosinophils and mast cells to induce bronchoconstriction, airway oedema, and mucus secretion. Airway wall remodelling changes include an increase in the thickness of the airway smooth muscle with hypertrophy and hyperplasia, thickening of the lamina reticularis with increased collagen and tenascin deposition, an increase in blood vessels and goblet cell numbers in the airway epithelium [6].

COPD: definition and pathophysiology

COPD is a preventable and treatable disease with some significant extrapulmonary effects that may contribute to the severity in individual patients and its pulmonary component is characterised by airflow limitation that is not fully reversible, usually progressive and associated with an abnormal inflammatory response of the lung to noxious particles or gases (http://www.goldcopd.com). The characteristic pathology of stable COPD includes chronic inflammation of the small bronchi and bronchioles less than 2 mm diameter, together with structural changes to all of the airways and lungs with emphysema, with glandular hypertrophy and hyperplasia with goblet cell hyperplasia, cellular inflammation, increased airway smooth muscle mass and distortion due to fibrosis of bronchioles. The chronic inflammation is characterised by an accumulation of neutrophils, macrophages, B cells, lymphoid aggregates and CD8$^+$ T cells [7] and the degree of inflammation increases with the severity of disease [8]. Although inflammatory cells contribute to the volume of the tissue wall in the small airways in COPD, structural changes such as epithelial metaplasia increase in airway smooth muscle, goblet cell hyperplasia and submucosal gland hypertrophy are other constituents of this thickening which also increase with severity [8]. The degree of airflow limitation as measured by FEV$_1$ is also correlated with the degree of airway wall thickness, providing indirect evidence for a role for airway wall remodelling in airflow obstruction of COPD. Emphysema may also contribute to airflow limitation resulting from loss of lung elastic recoil [9].

The basic abnormality in COPD is the response of airway cells particularly epithelial cells and macrophages to toxic gases and particulates in cigarette smoke that generates an inflammatory and immune response with the release of cytokines, chemokines, growth factors and proteases [10]. In COPD, as contrasted to healthy smokers also exposed to cigarette smoke, these mechanisms may be amplified by the effect of oxidative stress, perhaps viruses and by unknown genetic factors. The inflammatory response includes an excessive recruitment of neutrophils and immune cells including CD8 cytotoxic T cells. There is increasing elastolysis with the involvement of serine proteases, cathepsins and matrix metalloproteinase leading to alveolar wall destruction (emphysema). The chronic inflammatory process together with airway wall remodelling affects particularly the small airways. Within these mechanisms, airway smooth muscle cells are involved and may contribute particularly to the processes occurring in the small airways.

MMPs in the airways, including airway smooth muscle

Many inflammatory cells including macrophages, neutrophils, eosinophils, T cells and resident cells such as fibroblasts, epithelial cells and airway smooth muscle cells in the airways are capable of synthesising and releasing MMPs. The majority of MMPs are not expressed in normal healthy tissues but are expressed in diseased tissues that are inflamed or undergoing repair and remodelling [11]. MMP expression may be upregulated by exogenous stimuli, cytokines and cell–cell contact. Conversely, cytokines such as interferon (IFN)-γ and interleukins (IL)-4 and -10 may downregulate MMP expression. Both inflammatory and stromal cells can express MMPs, although the profile is both cell and stimulus specific.

Bronchial epithelial cells release MMP-2, MMP-9 and their major inhibitor, TIMP-1. Pulmonary epithelial cells may be a significant source of MMPs as they express MMP-1, 2, 7 and 9 [12–14]. Alveolar epithelial cells and Type II pneumocytes, produce TIMP-2 and MMP-1 [15]. Tumour necrosis factor (TNF)-α, EGF and IFN-γ increase gene and protein expression of MMP-12 in human bronchial epithelial cells [16]. Macrophages express a wider profile and greater quantities of MMPs than monocytes [17]. Neutrophils can release MMP-8 and MMP-9 on activation [18–20].

Airway smooth muscle cells is also an important source of MMPs and may therefore be involved in asthma and COPD as an important source of MMPs. Human airway smooth muscle cells express MMP-1, MMP-2, MMP-3, MMP-9 and MMP-14 [21–24]. MMP-3 is usually bound to airway smooth muscle-derived matrix, consistent with the staining for MMP-3 in the submucosal matrix of patients with chronic asthma [25]. IL-1β activates human airway smooth muscle cells to secrete MMP-9 [26] and MMP-12 [27]. Active MMP-12 is present in airway smooth muscle cells of small airways of smokers and COPD patients [28]. In addition to acting as a gelatinase where exatracellular matrix may be degraded, and contribute to emphysema (close proximity of airway smooth muscle bundle in small airways to emphysematous lung), MMP-12 also releases TNF-α from pro-TNF-α [29].

Airway smooth muscle cells also produce tissue inhibitor of metalloproteinase (TIMP) particularly TIMP-1 and TIMP-2 which counteract the proteolytic activity of secreted MMPs [21]. Whether there is a shift in the MMP/TIMP balance in airway smooth muscle cells in airways disease remains to be determined. However, there is evidence to indicate that MMPs may contribute to airway wall remodelling by modulating airway smooth muscle proliferation and migration. MMPs regulate airway smooth muscle hyperplasia by causing the release of immobilised growth factors, such as the release of transforming growth factor (TGF)-β when the extracellualr matrix (ECM) proteoglycan, decorin, is degraded by MMPs [30]. In addition, MMPs may degrade insulin-like growth factor binding proteins causing the release of insulin-growth factor (IGF) [31]. IGF-II is released from airway smooth muscle cells and induces airway smooth muscle proliferation [32]. The cells in the lung are constantly exposed to forces of stretch and relaxation and excessive stretch in airway disease due to excessive airway narrowing may potentially contribute to airway wall remodelling by promoting airway smooth muscle cell proliferation and migration. Mechanical strain applied to human airway smooth muscle cells in culture increases their proliferation and migration by inducing expression of extracellular MMP inducer (EMMPRIN) which can lead to the subsequent release and activation of MMP-1, 2, 3 and MT1-MMP [33].

MMP-12 may facilitate airway inflammation by stimulating migration of inflammatory cells such as monocytes and macrophages to inflammatory sites, and mediate airway remodelling by degrading ECM proteins through its enzymatic activity or through mediating inflammatory cytokines to induce other MMPs, including MMP-2, 9, 13 and 14, in the lung [34]. Overproduction of MMP-12 causes pathological ECM protein breakdown and excessive airway remodelling, which has been implicated in asthma and COPD. Studies from MMP-12 knockout mice indicate that MMP-12 is a key mediator in cigarette smoke-induced emphysema [35]. Xie and co-workers provide evidence that MMP-12 mRNA and protein are expressed by *in situ* human airway smooth muscle cells obtained from bronchial biopsies of normal volunteers, and of patients with asthma, COPD and chronic cough [28]. The pro-inflammatory cytokine, interleukin (IL)-1β, induced a >100-fold increase in MMP-12 gene expression and a >10-fold enhancement in MMP-12 activity of primary airway smooth muscle cell cultures. Selective inhibitors of extracellular signal-regulated kinase, c-Jun N-terminal kinase and phosphatidylinositol 3-kinase reduced the activity of IL-1β on MMP-12, indicating a role for these kinases in IL-1β-induced induction and release of MMP-12. IL-1β-induced MMP-12 activity and gene expression was downregulated by the corticosteroid dexamethasone but upregulated by the inflammatory cytokine TNF-α through enhancing activator protein-1 activation by IL-1β. TGF-β had no significant effect on MMP-12 induction [28].

MMPs in asthma

MMP-9 was the first MMP to be implicated in the pathology of asthma and about which there is most data. Increased levels of active MMP-9 have been detected in bronchoalveolar lavage fluid [36], and in sputum [37, 38] and in serum [39]. In sputum, 60% of asthmatics had detectable amounts of the 85 kDa activated form of MMP-9, but was not found in sputum from control subjects. In this study, MMP-2 and also higher levels of TIMP-1 has been reported [37]. More recently, increased mRNA transcripts for MMP-1 and TIMP-1 in sputum cell pellets containing inflammatory and epithelial cells from asthmatics compared to controls was reported [40]. Increased MMP-9 activity was correlated with the increased number of neutrophils in the airways [41, 42]. One major source of MMP-9 has been the eosinophil [43]. However, MMP-9 has also been shown to be expressed in the bronchial epithelium, and its expression was correlated with the eosinophil counts, linking MMP-9 to eosinophilic infiltration in the airways [43, 44]. The level of MMP-9 in sputum or in serum have been related to the severity of asthma, with more severe asthmatics showing higher levels than those with mild asthma [45, 46]. In the more severe category of asthma, Wenzel et al. found a significantly greater proportion of patients with severe asthma demonstrated MMP-9 immunostaining of the subepithelial basement membrane than control subjects with raised levels in BAL fluid [47]. The high molecular weight form of MMP-9 was found to be significantly higher in BAL fluid from patients with severe asthma than moderate asthma, and a strong correlation between MMP-9 and the number of neutrophils in BAL fluid was found [48].

This concept of severity of asthma linked to higher levels of MMP-9 is supported by results from studies of spontaneous episodes of asthma worsening. Thus, sputum MMP-9 levels and activity of MMP-9 were significantly elevated in asthmatic patients during an acute episode than in stable asthmatic patients; this elevation decreased after 7 and 28 days of therapy with corticosteroids, theophylline and salbutamol [49]. Plairway smooth musclea concentrations and activity of MMP-9 were also increased during an acute severe asthma attack, with no significant changes in MMP-2 or TIMP-1 levels [46, 50]. Similarly, in nocturnal asthma, BAL levels of MMP-9 and the MMP-9/TIMP-1 ratio both increase at night [51]. In acute severe asthma patients that are being ventilated, the 92 KDa MMP-9 concentration was increased 10–160-fold in epithelial lining fluid, together with increases in the activated form of MMP-3.

Finally, allergen exposure models in asthmatics show that MMP-9 is released. Thus, MMP-9 levels were increased in sputum from allergic asthmatics exposed to whole lung allergen exposure, and this correlated with the eosinophil percentage counts [18]. Levels of TIMP-1 did not increase. A correlation was observed between MMP-9 levels and the maximum fall in FEV1 during the late response, and the number of neutrophils in sputum [52]. Segmental allergen challenge also resulted in an increase in MMP-9 in the airways 48 h after allergen challenge [53].

Against this background, an immunological study of total MMP-3 and MMP-9 immunolocalisation in bronchial biopsies showed that MMP-9 was detectable within the extracellular matrix and was co-localised to neutrophils, while MMP-3 was co-localised to mast cells, eosinophils and neutrophils, epithelium and the lamina reticularis. However, there was no difference in the extent of matrix immunoreactivity for either MMP between mild or severe asthma and nonasthma controls [25].

MMPs in COPD

Activities of MMP-9 have been measured in sputum and BAL fluid of patients with COPD and found to be elevated [37, 54]. Increased gelatinolytic activity in sputum due to MMP-2 and MMP-9 activities has been reported with 85% of COPD patients but none of the controls expressed the activated form of MMP-9 (85 kDa). Pro-MMP-2 was found in 25% of COPD patients with only 5% of controls. Levels of TIMP-1 were higher in COPD patients than in controls [18]. Similar data were reported by Beeh et al. [54]. Increased amounts of collagenase activity was also reported in BAL fluid from COPD patients, likely to be due to activity of MMP-8 compared to healthy controls. MMP-9 was present in the majority of COPD patients [55]. Another study indicated that the rise in MMP-8 and MMP-9 in smokers was associated with the presence of emphysema [56]. Alveolar macrophages obtained from normal smokers express more MMP-9 than those from normal subjects [57] and there is an even greater increase in cells from COPD patients that release greater amounts of MMP-9 with greater enzymatic activity than do alveolar macrophages from healthy smokers and non-smokers [58].

Expression of MMPs have also been investigated in lung tissue from COPD patients. An increased immunostaining expression of MMP-1, 2, 8 and 9 has been observed in lung tissue of COPD patients. They reported MMP-1 and MMP-2 localised to alveolar macrophages and interstitial macrophages and in epithelial cells, and MMP-8 and MMP-9 primarily in neutrophils [59]. Expression of MMPs have also been looked at in relationship to the emphysema of COPD. Increased MMP-1 expression is reported in lung parenchyma of emphysema patients, but there are no significant differences in MMP-9 expression from emphysema patients when compared with controls [60]. However, another study that examined protein extracts from human lung tissues found that MMP-9 levels and MMP-9/TIMP-1 ratio were correlated with the amount of cigarette smoking. MMP-9 concentrations were inversely correlated with FEV_1 [61]. Concerning MMP-9, the frequency of the -1562C/T polymorphisms in its promoter region (associated with a higher promoter activity than the C allele) was found to be higher in smokers with emphysema compared to those without emphysema [62].

Studies in transgenic expression of human MMP-1 in lungs of mice showed a disruption of the lung architecture with emphysema [63], associated with a marked increase in Type III collagen, a vital structural element of the alveolar walls. On the other hand, in a murine model of asthma, deficiency in MMP-9 resulted in decreased lymphocytic inflammation and peribronchial mononuclear cell infiltration, indicating the potential role of MMP-9 in cell chemotaxis. MMP-12 levels are also enhanced in BAL fluid of COPD patients. The number of MMP-12 expressing macrophages in BAL fluid and in tissue sections of bronchial biopsies was found to be higher in COPD patients [64]. Increased levels and activity of MMP-12 has also been reported in sputum of patients with COPD compared to healthy controls or smokers [65]. Increased elastolytic activity in macrophages from patients with COPD caused by exposure to wood smoke has been reported and this enzymatic activity was related to MMP-12 [66]. MMP-2 and MMP-12, but not MMP-9, gene expression were increased in COPD patients. Polymorphisms of MMP-12 gene but not those of MMP-9 have been associated with the rate of decline of lung function in smokers [67].

MMPs may play a major role in matrix turnover, remodelling and angiogenesis in COPD [68]. Overexpression of IL-1β in mouse lungs induces inflammatory and morphological changes consistent with a COPD phenotype. Of note, the inflammatory response is characterised by increased production of MMP-9 and MMP-12 [69]. The importance of MMP-9 and MMP-12 in the development of emphysema has been demonstrated in MMP-12 knockout mice which are completely protected against the development of smoke-induced emphysema [35] and in a more recent study where a chemical MMP-9 and MMP-12 inhibitor prevented the development of smoke-induced emphysema and small airways remodelling in guinea pigs [70].

MMP-12, also called macrophage metalloelastase, was originally detected in alveolar macrophages of cigarette smokers [71]. It is secreted as a 54 kDa inactive pro-enzyme which is activated by proteolytic cleavage of the prodomain followed by processing into two active enzymes of 45 kDa and 22 kDa [71]. MMP-12 degrades a broad range of ECM proteins, including elastin, type IV collagen, fibronectin, laminin and gelatine [72, 73], and is involved in turnover of the matrix, cell migration, tissue repairing and remodelling. In addition, MMP-12 can activate other MMPs, for example, MMP-2 and 3, leading to subsequent degradation of other ECM proteins [74]. MMP-12 is also produced by normal human bronchial epithelial cells and airway smooth muscle cells [16]. Mice exposed to cigarette smoke express more MMP-12 mRNA than non-exposed mice [75]. MMP-12 gene deletion in mice protects from the development of emphysema after long-term exposure to cigarette smoke [35]. In mouse models of the acute response to tobacco smoke, the secretion of MMP-12 from macrophages has been reported to cause macrophages to release TNF-α, resulting in neutrophil influx, possibly through activation of vascular endothelial cells [76].

Role of MMP activation in asthma and COPD

MMPs may participate in the pathophysiology of asthma and COPD in many ways. Apart from the obvious degradation of extracellular matrix proteins and destruction of lung tissue, the potential actions include many aspects involved with the regulation of airway inflammation such as the extravasation of inflammatory cells from the blood compartment to the extracellular space, the movement of inflammatory cells through the epithelium, and the chemotaxis of inflammatory cells into the inflamed tissue by modulating the activities of chemoattractants.

Interactions with Th2 cytokines

IL-13 effects on airway wall remodelling and induction of emphysema may be related to overproduction of MMPs. IL-13 dependent expression of MMP-12 is required for the development of allergic airway eosinophilia in mice [77]. In a murine model of cigarette smoking exposure, IL-13 overexpression induces MMP-2, MMP-9, MMP-12, MMP-13 and MMP-14 expression [78, 79]. Blocking MMPs led to a decrease in IL-13 induced emphysema, and MMP-9 and MMP-12 dependent pathways were particularly involved in these IL-13-dependent effects [80]. IL-13 overexpression induces lung fibrosis, the effect of which was shown to be through TGF-β expression [79], which in turn may be activated by MMP-2, MMP-9 and MMP-12 [81–83].

Interactions with chemokines

MMPs may also be involved in chemotaxis of inflammatory cells by maintaining gradients of chemokines or cytokines or interacting directly with chemokines or cytokines. MMPs may convert chemokines into antagonist molecules and therefore act as an anti-inflammatory. Thus, MMP-1, MMP-3, MMP-13 and MMP-14 cleaves the N-terminal portion of the following chemokines that may be involved in asthma and COPD, namely CCL2/MCP 1, CCL8/MCP 2 and CCL13/MCP 4 to produce antagonistic factors [84]. On the other hand, MMP-1, 2, 3, 9, 13 and 14 activate CXCL11/SDF 1, while MMP-9 can process CXCL5/ENA 78, CXCL6/GCP 2 to inactivate them, but it increases the activity of IL-8/CXCL8 [85, 86]. MMP-9 also is required for transepithelial gradients of CCL11/eotaxin, facilitating the movement of chemokines from one compartment to another [87]. MMP-2-deficient mice demonstrate a reduced degree of allergic inflammation due to decreased levels of CCL11/eotaxin [88]. On the other hand, MMP-12 is required for the influx of macrophages to the lungs of cigarette-smoke exposed lungs [35]; this effect may be indirect through the effect of MMP-12 in generating chemotactic fragments from elastin. Elastin fragments in bronchoalveolar lavage fluid are chemotactic for macrophages; a monoclonal antibody against elastin fragments inhibited this chemotactic activity and mice lacking MMP-12 did not have elastin fragments [89].

Interactions with growth factors

MMPs also can directly or indirectly affect the activity of cytokines involved in inflammation and airway wall remodelling, such as VEGF and EGF [90, 91]. VEGF signalling may regulate MMP-9 expression in an ovalbumin murine model [92]. Latent TGF-β1 can be activated by MMP-3, MMP-9 and MMP-12 [81–83]. MMP-9 in turn could then modulate TGF-β1 [93].

MMP-1 secreted by hyperplastic smooth muscle cells activates IGF axis [94]. IGF-1 is produced by airway epithelial cells and induces airway smooth muscle cell proliferation through its ability to upregulate MMP-2, an airway smooth muscle mitogen [95]. Inhibition of MMP-2 reduces smooth muscle proliferation [24].

Conclusion

There is little doubt that MMPs are involved in the pathophysiology of asthma and COPD, both in the inflammatory and remodelling processes that are prominent in these conditions. However, their contribution is potentially complex since MMPs may have both beneficial as well as deleterious effects. Thus, some actions may lead to pro-inflammatory effects, while others may also cause anti-inflammatory effects.

Interference with the effects of certain MMPs may lead to reduced destruction of the lungs while development of fibrosis may also result. Therefore, the precise role of these MMPs need to be clarified carefully before selective MMP inhibitors are tried for therapeutic aims in these diseases.

References

1   Gross J, Lapiere CM (1962) Collagenolytic activity in amphibian tissues: a tissue culture assay. *Proc Natl Acad Sci USA* 48: 1014–1022

2   Brinckerhoff CE, Matrisian LM (2002) Matrix metalloproteinases: a tail of a frog that became a prince. *Nat Rev Mol Cell Biol* 3: 207–214

3   Davidson JM (1990) Biochemistry and turnover of lung interstitium. *Eur Respir J* 3: 1048–1063

4   Van Wart HE, Birkedal-Hansen H (1990) The cysteine switch: a principle of regulation of metalloproteinase activity with potential applicability to the entire matrix metallo-proteinase gene family. *Proc Natl Acad Sci USA* 87: 5578–5582

5   Brew K, Dinakarpandian D, Nagase H (2000) Tissue inhibitors of metalloproteinases: evolution, structure and function. *Biochim Biophys Acta* 1477: 267–283

6   Bousquet J, Jeffery PK, Busse WW, Johnson M, Vignola AM (2000) Asthma. From bronchoconstriction to airways inflammation and remodeling. *Am J Respir Crit Care Med* 161: 1720–1745

7   Hogg JC (2004) Pathophysiology of airflow limitation in chronic obstructive pulmonary disease. *Lancet* 364: 709–721

8   Hogg JC, Chu F, Utokaparch S, Woods R, Elliott WM, Buzatu L, Cherniack RM, Rogers RM, Sciurba FC, Coxson HO et al. (2004) The nature of small-airway obstruction in chronic obstructive pulmonary disease. *N Engl J Med* 350: 2645–2653

9   Kim WD, Eidelman DH, Izquierdo JL, Ghezzo H, Saetta MP, Cosio MG (1991) Centrilobular and panlobular emphysema in smokers. Two distinct morphologic and functional entities. *Am Rev Respir Dis* 144: 1385–1390

10 Barnes PJ, Shapiro SD, Pauwels RA (2003) Chronic obstructive pulmonary disease: molecular and cellular mechanisms. *Eur Respir J* 22: 672–688

11 Parks WC, Shapiro SD (2001) Matrix metalloproteinases in lung biology. *Respir Res* 2: 10–19

12 Yao PM, Buhler JM, d'Ortho MP, Lebargy F, Delclaux C, Harf A, Lafuma C (1996) Expression of matrix metalloproteinase gelatinases A and B by cultured epithelial cells from human bronchial explants. *J Biol Chem* 271: 15580–15589

13 Dunsmore SE, Saarialho-Kere UK, Roby JD, Wilson CL, Matrisian LM, Welgus HG, Parks WC (1998) Matrilysin expression and function in airway epithelium. *J Clin Invest* 102: 1321–1331

14 Mercer BA, Kolesnikova N, Sonett J, D'Armiento J (2004) Extracellular regulated kinase/mitogen activated protein kinase is up-regulated in pulmonary emphysema and

mediates matrix metalloproteinase-1 induction by cigarette smoke. *J Biol Chem* 279: 17690–17696

15 Xu J, Benyon RC, Leir SH, Zhang S, Holgate ST, Lackie PM (2002) Matrix metalloproteinase-2 from bronchial epithelial cells induces the proliferation of subepithelial fibroblasts. *Clin Exp Allergy* 32: 881–888

16 Lavigne MC, Thakker P, Gunn J, Wong A, Miyashiro JS, Wasserman AM, Wei SQ, Pelker JW, Kobayashi M, Eppihimer MJ (2004) Human bronchial epithelial cells express and secrete MMP-12. *Biochem Biophys Res Commun* 324: 534–546

17 Campbell EJ, Cury JD, Shapiro SD, Goldberg GI, Welgus HG (1991) Neutral proteinases of human mononuclear phagocytes. Cellular differentiation markedly alters cell phenotype for serine proteinases, metalloproteinases, and tissue inhibitor of metalloproteinases. *J Immunol* 146: 1286–1293

18 Cataldo DD, Bettiol J, Noel A, Bartsch P, Foidart JM, Louis R (2002) Matrix metalloproteinase-9, but not tissue inhibitor of matrix metalloproteinase-1, increases in the sputum from allergic asthmatic patients after allergen challenge. *Chest* 122: 1553–1559

19 Takafuji S, Ishida A, Miyakuni Y, Nakagawa T (2003) Matrix metalloproteinase-9 release from human leukocytes. *J Investig Allergol Clin Immunol* 13: 50–55

20 Claesson R, Johansson A, Belibasakis G, Hanstrom L, Kalfas S (2002) Release and activation of matrix metalloproteinase 8 from human neutrophils triggered by the leukotoxin of *Actinobacillus actinomycetemcomitans. J Periodontal Res* 37: 353–359

21 Elshaw SR, Henderson N, Knox AJ, Watson SA, Buttle DJ, Johnson SR (2004) Matrix metalloproteinase expression and activity in human airway smooth muscle cells. *Br J Pharmacol* 142: 1318–1324

22 Rajah R, Nunn SE, Herrick DJ, Grunstein MM, Cohen P (1996) Leukotriene D4 induces MMP-1, which functions as an IGFBP protease in human airway smooth muscle cells. *Am J Physiol* 271: L1014–L1022

23 Foda HD, George S, Rollo E, Drews M, Conner C, Cao J, Panettieri Jr RA, Zucker S (1999) Regulation of gelatinases in human airway smooth muscle cells: mechanism of progelatinase A activation. *Am J Physiol* 277: L174–L182

24 Johnson S, Knox A (1999) Autocrine production of matrix metalloproteinase-2 is required for human airway smooth muscle proliferation. *Am J Physiol* 277: L1109–L1117

25 Dahlen B, Shute J, Howarth P (1999) Immunohistochemical localisation of the matrix metalloproteinases MMP-3 and MMP-9 within the airways in asthma. *Thorax* 54: 590–596

26 Liang KC, Lee CW, Lin WN, Lin CC, Wu CB, Luo SF, Yang CM (2007) Interleukin-1beta induces MMP-9 expression *via* p42/p44 MAPK, p38 MAPK, JNK, and nuclear factor-kappaB signaling pathways in human tracheal smooth muscle cells. *J Cell Physiol* 211: 759–770

27 Xie S, Sukkar MB, Issa R, Oltmanns U, Nicholson AG, Chung KF (2005) Regulation of TGF-β1-induced connective tissue growth factor expression in airway smooth muscle cells. *Am J Physiol Lung Cell Mol Physiol* 288: L68–L76

28 Xie S, Issa R, Sukkar MB, Oltmanns U, Bhavsar PK, Papi A, Caramori G, Adcock I, Chung KF (2005) Induction and regulation of matrix metalloproteinase-12 in human airway smooth muscle cells. *Respir Res* 6: 148

29 Chandler S, Cossins J, Lury J, Wells G (1996) Macrophage metalloelastase degrades matrix and myelin proteins and processes a tumour necrosis factor-alpha fusion protein. *Biochem Biophys Res Commun* 228: 421–429

30 Imai K, Hiramatsu A, Fukushima D, Pierschbacher MD, Okada Y (1997) Degradation of decorin by matrix metalloproteinases: identification of the cleavage sites, kinetic analyses and transforming growth factor-beta1 release. *Biochem J* 322 (Pt 3): 809–814

31 Fowlkes JL, Enghild JJ, Suzuki K, Nagase H (1994) Matrix metalloproteinases degrade insulin-like growth factor-binding protein-3 in dermal fibroblast cultures. *J Biol Chem* 269: 25742–25746

32 Noveral JP, Bhala A, Hintz RL, Grunstein MM, Cohen P (1994) Insulin-like growth factor axis in airway smooth muscle cells. *Am J Physiol* 267: L761–L765

33 Hasaneen NA, Zucker S, Cao J, Chiarelli C, Panettieri RA, Foda HD (2005) Cyclic mechanical strain-induced proliferation and migration of human airway smooth muscle cells: role of EMMPRIN and MMPs. *FASEB J* 19:1507–1509

34 Lanone S, Zheng T, Zhu Z, Liu W, Lee CG, Ma B, Chen Q, Homer RJ, Wang J, Rabach LA et al (2002) Overlapping and enzyme-specific contributions of matrix metalloproteinases-9 and -12 in IL-13-induced inflammation and remodeling. *J Clin Invest* 110: 463–474

35 Hautamaki RD, Kobayashi DK, Senior RM, Shapiro SD (1997) Requirement for macrophage elastase for cigarette smoke-induced emphysema in mice. *Science* 277: 2002–2004

36 Mautino G, Oliver N, Chanez P, Bousquet J, Capony F (1997) Increased release of matrix metalloproteinase-9 in bronchoalveolar lavage fluid and by alveolar macrophages of asthmatics. *Am J Respir Cell Mol Biol* 17: 583–591

37 Cataldo D, Munaut C, Noel A, Frankenne F, Bartsch P, Foidart JM, Louis R (2000) MMP-2- and MMP-9-linked gelatinolytic activity in the sputum from patients with asthma and chronic obstructive pulmonary disease. *Int Arch Allergy Immunol* 123: 259–267

38 Vignola AM, Riccobono L, Mirabella A, Profita M, Chanez P, Bellia V, Mautino G, D'accardi P, Bousquet J, Bonsignore G (1998) Sputum metalloproteinase-9/tissue inhibitor of metalloproteinase-1 ratio correlates with airflow obstruction in asthma and chronic bronchitis. *Am J Respir Crit Care Med* 158: 1945–1950

39 Bosse M, Chakir J, Rouabhia M, Boulet LP, Audette M, Laviolette M (1999) Serum matrix metalloproteinase-9: Tissue inhibitor of metalloproteinase-1 ratio correlates with steroid responsiveness in moderate to severe asthma. *Am J Respir Crit Care Med* 159: 596–602

40 Cataldo DD, Gueders M, Munaut C, Rocks N, Bartsch P, Foidart JM, Noel A, Louis R (2004) Matrix metalloproteinases and tissue inhibitors of matrix metalloproteinases mRNA transcripts in the bronchial secretions of asthmatics. *Lab Invest* 84: 418–424

41   Cataldo D, Munaut C, Noel A, Frankenne F, Bartsch P, Foidart JM, Louis R (2001) Matrix metalloproteinases and TIMP-1 production by peripheral blood granulocytes from COPD patients and asthmatics. *Allergy* 56: 145–151

42   Prause O, Bozinovski S, Anderson GP, Linden A (2004) Increased matrix metalloproteinase-9 concentration and activity after stimulation with interleukin-17 in mouse airways. *Thorax* 59: 313–317

43   Ohno I, Ohtani H, Nitta Y, Suzuki J, Hoshi H, Honma M, Isoyama S, Tanno Y, Tamura G, Yamauchi K et al (1997) Eosinophils as a source of matrix metalloproteinase-9 in asthmatic airway inflammation. *Amer J Respir Cell Mol Biol* 16: 212–219

44   Han Z, Junxu, Zhong N (2003) Expression of matrix metalloproteinases MMP-9 within the airways in asthma. *Respir Med* 97: 563–567

45   Mattos W, Lim S, Russell R, Jatakanon A, Chung KF, Barnes PJ (2002) Matrix metalloproteinase-9 expression in asthma: effect of asthma severity, allergen challenge, and inhaled corticosteroids. *Chest* 122: 1543–1552

46   Belleguic C, Corbel M, Germain N, Lena H, Boichot E, Delaval PH, Lagente V (2002) Increased release of matrix metalloproteinase-9 in the plairway smooth musclea of acute severe asthmatic patients. *Clin Exp Allergy* 32: 217–223

47   Wenzel SE, Balzar S, Cundall M, Chu HW (2003) Subepithelial basement membrane immunoreactivity for matrix metalloproteinase 9: association with asthma severity, neutrophilic inflammation, and wound repair. *J Allergy Clin Immunol* 111: 1345–1352

48   Cundall M, Sun Y, Miranda C, Trudeau JB, Barnes S, Wenzel SE (2003) Neutrophil-derived matrix metalloproteinase-9 is increased in severe asthma and poorly inhibited by glucocorticoids. *J Allergy Clin Immunol* 112: 1064–1071

49   Lee YC, Lee HB, Rhee YK, Song CH (2001) The involvement of matrix metalloproteinase-9 in airway inflammation of patients with acute asthma. *Clin Exp Allergy* 31: 1623–1630

50   Oshita Y, Koga T, Kamimura T, Matsuo K, Rikimaru T, Aizawa H (2003) Increased circulating 92 kDa matrix metalloproteinase (MMP-9) activity in exacerbations of asthma. *Thorax* 58: 757–760

51   Pham DN, Chu HW, Martin RJ, Kraft M (2003) Increased matrix metalloproteinase-9 with elastolysis in nocturnal asthma. *Ann Allergy Asthma Immunol* 90: 72–78

52   Boulay ME, Prince P, Deschesnes F, Chakir J, Boulet LP (2004) Metalloproteinase-9 in induced sputum correlates with the severity of the late allergen-induced asthmatic response. *Resp* 71: 216–224

53   Kelly EA, Busse WW, Jarjour NN (2000) Increased matrix metalloproteinase-9 in the airway after allergen challenge. *Am J Respir Crit Care Med* 162: 1157–1161

54   Beeh KM, Beier J, Kornmann O, Buhl R (2003) Sputum matrix metalloproteinase-9, tissue inhibitor of metalloprotinease-1, and their molar ratio in patients with chronic obstructive pulmonary disease, idiopathic pulmonary fibrosis and healthy subjects. *Respir Med* 97: 634–639

55   Finlay GA, O'Driscoll LR, Russell KJ, D'Arcy EM, Masterson JB, FitzGerald MX,

O'Connor CM (1997) Matrix metalloproteinase expression and production by alveolar macrophages in emphysema. *Am J Respir Crit Care Med* 156: 240–247

56   Betsuyaku T, Nishimura M, Takeyabu K, Tanino M, Venge P, Xu S, Kawakami Y (1999) Neutrophil granule proteins in bronchoalveolar lavage fluid from subjects with subclinical emphysema. *Am J Respir Crit Care Med* 159: 1985–1991

57   Lim S, Roche N, Oliver BG, Mattos W, Barnes PJ, Chung KF (2000) Balance of matrix metalloprotease-9 and tissue inhibitor of metalloprotease-1 from alveolar macrophages in cigarette smokers. Regulation by interleukin-10. *Am J Respir Crit Care Med* 162: 1355–1360

58   Russell RE, Culpitt SV, DeMatos C, Donnelly L, Smith M, Wiggins J, Barnes PJ (2002) Release and activity of matrix metalloproteinase-9 and tissue inhibitor of metalloproteinase-1 by alveolar macrophages from patients with chronic obstructive pulmonary disease. *Amer J Respir Cell Mol Biol* 26: 602–609

59   Segura-Valdez L, Pardo A, Gaxiola M, Uhal BD, Becerril C, Selman M (2000) Upregulation of gelatinases A and B, collagenases 1 and 2, and increased parenchymal cell death in COPD. *Chest* 117: 684–694

60   Imai K, Dalal SS, Chen ES, Downey R, Schulman LL, Ginsburg M, D'Armiento J (2001) Human collagenase (matrix metalloproteinase-1) expression in the lungs of patients with emphysema. *Am J Respir Crit Care Med* 163: 786–791

61   Kang MJ, Oh YM, Lee JC, Kim DG, Park MJ, Lee MG, Hyun IG, Han SK, Shim YS, Jung KS (2003) Lung matrix metalloproteinase-9 correlates with cigarette smoking and obstruction of airflow. *J Korean Med Sci* 18: 821–827

62   Minematsu N, Nakamura H, Tateno H, Nakajima T, Yamaguchi K (2001) Genetic polymorphism in matrix metalloproteinase-9 and pulmonary emphysema. *Biochem Biophys Res Commun* 289: 116–119

63   D'Armiento J, Dalal SS, Okada Y, Berg RA, Chada K (1992) Collagenase expression in the lungs of transgenic mice causes pulmonary emphysema. *Cell* 71: 955–961

64   Molet S, Belleguic C, Lena H, Germain N, Bertrand CP, Shapiro SD, Planquois JM, Delaval P, Lagente V (2005) Increase in macrophage elastase (MMP-12) in lungs from patients with chronic obstructive pulmonary disease. *Inflamm Res* 54: 31–36

65   Demedts IK, Morel-Montero A, Lebecque S, Pacheco Y, Cataldo D, Joos GF, Pauwels RA, Brusselle GG (2006) Elevated MMP-12 protein levels in induced sputum from patients with COPD. *Thorax* 61: 196–201

66   Montano M, Beccerril C, Ruiz V, Ramos C, Sansores RH, Gonzalez-Avila G (2004) Matrix metalloproteinases activity in COPD associated with wood smoke. *Chest* 125: 466–472

67   Joos L, He JQ, Shepherdson MB, Connett JE, Anthonisen NR, Pare PD, Sandford AJ (2002) The role of matrix metalloproteinase polymorphisms in the rate of decline in lung function. *Hum Mol Genet* 11: 569–576

68   Shapiro SD (2002) Proteinases in chronic obstructive pulmonary disease. *Biochem Soc Trans* 30: 98–102

69 Lappalainen U, Whitsett JA, Wert SE, Tichelaar JW, Bry K (2005) Interleukin-1beta causes pulmonary inflammation, emphysema, and airway remodeling in the adult murine lung. *Am J Respir Cell Mol Biol* 32: 311–318

70 Churg A, Wang R, Wang X, Onnervik PO, Thim K, Wright JL (2007) Effect of an MMP-9/MMP-12 inhibitor on smoke-induced emphysema and airway remodelling in guinea pigs. *Thorax* 62: 706–713

71 Shapiro SD, Kobayashi DK, Ley TJ (1993) Cloning and characterization of a unique elastolytic metalloproteinase produced by human alveolar macrophages. *J Biol Chem* 268: 23824–23829

72 Chandler S, Cossins J, Lury J, Wells G (1996) Macrophage metalloelastase degrades matrix and myelin proteins and processes a tumour necrosis factor-alpha fusion protein. *Biochem Biophys Res Commun* 228: 421–429

73 Gronski TJ Jr, Martin RL, Kobayashi DK, Walsh BC, Holman MC, Huber M, Van Wart HE, Shapiro SD (1997) Hydrolysis of a broad spectrum of extracellular matrix proteins by human macrophage elastase. *J Biol Chem* 272: 12189–12194

74 Matsumoto S, Kobayashi T, Katoh M, Saito S, Ikeda Y, Kobori M, Masuho Y, Watanabe T (1998) Expression and localization of matrix metalloproteinase-12 in the aorta of cholesterol-fed rabbits: relationship to lesion development. *Am J Pathol* 153: 109–119

75 Bracke K, Cataldo D, Maes T, Gueders M, Noel A, Foidart JM, Brusselle G, Pauwels RA (2005) Matrix metalloproteinase-12 and cathepsin D expression in pulmonary macrophages and dendritic cells of cigarette smoke-exposed mice. *Int Arch Allergy Immunol* 138: 169–179

76 Churg A, Wang RD, Tai H, Wang X, Xie C, Dai J, Shapiro SD, Wright JL (2003) Macrophage metalloelastase mediates acute cigarette smoke-induced inflammation *via* tumor necrosis factor-alpha release. *Am J Respir Crit Care Med* 167: 1083–1089

77 Pouladi MA, Robbins CS, Swirski FK, Cundall M, McKenzie AN, Jordana M, Shapiro SD, Stampfli M (2004) Interleukin-13-dependent expression of matrix metalloproteinase-12 is required for the development of airway eosinophilia in mice. *Am J Respir Cell Mol Biol* 30: 84–90

78 Lanone S, Zheng T, Zhu Z, Liu W, Lee CG, Ma B, Chen Q, Homer RJ, Wang J, Rabach LA et al (2002) Overlapping and enzyme-specific contributions of matrix metalloproteinases-9 and -12 in IL-13-induced inflammation and remodeling. *J Clin Invest* 110: 463–474

79 Zheng T, Zhu Z, Wang Z, Homer RJ, Ma B, Riese RJ, Chapman HA, Shapiro SD, Elias JA (2000) Inducible targeting of IL-13 to the adult lung causes matrix metalloproteinase- and cathepsin-dependent emphysema. *J Clin Invest* 106: 1081–1093

80 Lee CG, Homer RJ, Zhu Z, Lanone S, Wang X, Koteliansky V, Shipley JM, Gotwals P, Noble P, Chen Q et al (2001) Interleukin-13 induces tissue fibrosis by selectively stimulating and activating transforming growth factor beta(1). *J Exp Med* 194: 809–821

81 Maeda S, Dean DD, Gomez R, Schwartz Z, Boyan BD (2002) The first stage of transforming growth factor beta1 activation is release of the large latent complex from

the extracellular matrix of growth plate chondrocytes by matrix vesicle stromelysin-1 (MMP-3). *Calcif Tissue Int* 70: 54–65

82 Yu Q, Stamenkovic I (2000) Cell surface-localized matrix metalloproteinase-9 proteolytically activates TGF-beta and promotes tumor invasion and angiogenesis. *Genes Dev* 14: 163–176

83 Karsdal MA, Larsen L, Engsig MT, Lou H, Ferreras M, Lochter A, Delaisse JM, Foged NT (2002) Matrix metalloproteinase-dependent activation of latent transforming growth factor-beta controls the conversion of osteoblasts into osteocytes by blocking osteoblast apoptosis. *J Biol Chem* 277: 44061–44067

84 McQuibban GA, Gong JH, Wong JP, Wallace JL, Clark-Lewis I, Overall CM (2002) Matrix metalloproteinase processing of monocyte chemoattractant proteins generates CC chemokine receptor antagonists with anti-inflammatory properties *in vivo*. *Blood* 100: 1160–1167

85 Van den Steen PE, Proost P, Wuyts A, Van Damme J, Opdenakker G (2000) Neutrophil gelatinase B potentiates interleukin-8 tenfold by aminoterminal processing, whereas it degrades CTAP-III, PF-4, and GRO-alpha and leaves RANTES and MCP-2 intact. *Blood* 96: 2673–2681

86 Van den Steen PE, Wuyts A, Husson SJ, Proost P, Van Damme J, Opdenakker G (2003) Gelatinase B/MMP-9 and neutrophil collagenase/MMP-8 process the chemokines human GCP-2/CXCL6, ENA-78/CXCL5 and mouse GCP-2/LIX and modulate their physiological activities. *Eur J Biochem* 270: 3739–3749

87 Corry DB, Kiss A, Song LZ, Song L, Xu J, Lee SH, Werb Z, Kheradmand F (2004) Overlapping and independent contributions of MMP2 and MMP9 to lung allergic inflammatory cell egression through decreased CC chemokines. *FASEB J* 18: 995–997

88 Corry DB, Rishi K, Kanellis J, Kiss A, Song LZ, Xu J, Feng L, Werb Z, Kheradmand F (2002) Decreased allergic lung inflammatory cell egression and increased susceptibility to asphyxiation in MMP2-deficiency. *Nat Immunol* 3: 347–353

89 Houghton AM, Quintero PA, Perkins DL, Kobayashi DK, Kelley DG, Marconcini LA, Mecham RP, Senior RM, Shapiro SD (2006) Elastin fragments drive disease progression in a murine model of emphysema. *J Clin Invest* 116: 753–759

90 Bergers G, Brekken R, McMahon G, Vu TH, Itoh T, Tamaki K, Tanzawa K, Thorpe P, Itohara S, Werb Z et al. (2000) Matrix metalloproteinase-9 triggers the angiogenic switch during carcinogenesis. *Nat Cell Biol* 2: 737–744

91 Suzuki M, Raab G, Moses MA, Fernandez CA, Klagsbrun M (1997) Matrix metalloproteinase-3 releases active heparin-binding EGF-like growth factor by cleavage at a specific juxtamembrane site. *J Biol Chem* 272: 31730–31737

92 Lee KS, Min KH, Kim SR, Park SJ, Park HS, Jin GY, Lee YC (2006) Vascular endothelial growth factor modulates matrix metalloproteinase-9 expression in asthma. *Am J Respir Crit Care Med* 174: 161–170

93 Santibanez JF, Guerrero J, Quintanilla M, Fabra A, Martinez J (2002) Transforming growth factor-beta1 modulates matrix metalloproteinase-9 production through the Ras/

MAPK signaling pathway in transformed keratinocytes. *Biochem Biophys Res Commun* 296: 267–273

94   Rajah R, Nachajon RV, Collins MH, Hakonarson H, Grunstein MM, Cohen P (1999) Elevated levels of the IGF-binding protein protease MMP-1 in asthmatic airway smooth muscle. *Amer J Respir Cell Mol Biol* 20: 199–208

95   Zhang D, Bar-Eli M, Meloche S, Brodt P (2004) Dual regulation of MMP-2 expression by the type 1 insulin-like growth factor receptor: the phosphatidylinositol 3-kinase/Akt and Raf/ERK pathways transmit opposing signals. *J Biol Chem* 279: 19683–19690

Role of matrix metalloproteases in pulmonary fibrosis

Annie Pardo[1] and Moisés Selman[2]

[1]Facultad de Ciencias, Universidad Nacional Autónoma de México, Ciudad Universitaria, CP 04510, México DF, México; [2]Instituto Nacional de Enfermedades Respiratorias, Tlalpan 4502, CP 14080, México DF, México

Abstract

Lung fibrosis is the final result of a large variety of stimuli including systemic and autoimmune reactions, exposure to organic and inorganic particles, drugs, and radiation. Independent of etiology, the fibrotic response in the lung can be visualized as a dynamic and highly integrated cellular response to persistent injury and may be related to a damage-triggered inflammatory response or to an aberrant epithelial or endothelial reaction. In any case, the key cellular mediator is the myofibroblast, which when activated is the major effector of the lung remodeling. Several matrix metalloproteases (MMPs) have been shown to participate in this pathological process. These enzymes play an essential but complex role in several interrelated processes that take place in the pathogenesis of lung fibrosis including extracellular matrix remodeling, basement membrane disruption, epithelial apoptosis, cell migration, and angiogenesis. This review will focus on the role of MMPs in the development of lung fibrosis.

Pulmonary fibrosis

Pulmonary fibrosis is the final result of a large and heterogeneous group of lung disorders, known as interstitial lung diseases (ILD). These disorders are also called diffuse infiltrative (or parenchymal) lung diseases because they affect not only the interstitial compartment but also the alveolar-capillary units, the alveolar spaces, and frequently the small airways (alveolar ducts, respiratory bronchioles, and terminal bronchioles) [1]. There are over 150 ILD that vary widely in etiology, clinicoradiologic presentation, histopathologic features, and clinical course. Occupational/environmental factors, drugs, some infections and systemic disorders, such as sarcoidosis and autoimmune rheumatic diseases are common causes of ILD. In addition, there is a subgroup of ILD of uncertain etiology known as idiopathic interstitial pneumonias from which, idiopathic pulmonary fibrosis with its histopathologic pattern of usual interstitial pneumonia (UIP) is the most common and certainly the most aggressive form of lung fibrosis [1, 2].

Matrix Metalloproteinases in Tissue Remodelling and Inflammation,
edited by Vincent Lagente and Elisabeth Boichot
© 2008 Birkhäuser Verlag Basel/Switzerland

The pathogenic mechanisms involved in the development of fibrosis are intricate but it has been recently proposed that, independent of etiology, ILD may follow at least two different cellular pathways for the development of lung fibrosis [3, 4].

In most ILD the fibrotic response is associated to a sustained and unresolved lung inflammation (*inflammatory pathway*). In this case, the cells involved in the alveolitis as well as some of the molecular mechanisms implicated in the exaggerated accumulation of extracellular matrix may differ from one to another ILD, but all antigen-presenting and inflammatory cell-types, i.e., macrophages and dendritic cells, lymphocytes, and polymorphonuclear cells have the potential to provoke a fibrotic reaction and several common chemokines, cytokines and growth factors are involved [5–8]. In the inflammatory pathway, up- or downregulation of chemokines and adhesion molecules play a critical in the appropriate regulation of the inflammatory/fibrotic response [9, 10]. These families of mediators are involved in leukocyte chemotaxis and migration through the endothelial barrier into the inflamed lungs. In a profibrotic scenario chemokines/chemokine receptors can enhance and perpetuate inflammation, alter angiogenesis, polarize the immune response to a T helper-2 like reaction and chemoattract bone marrow/circulating fibroblasts precursors.

By contrast, idiopathic pulmonary fibrosis (IPF), a pernicious lung disorder of unknown etiology, represents an inflammatory-independent epithelial-dependent fibrotic process mirroring an abnormal wound healing [3, 4, 11, 12]. In this case (*epithelial pathway*), aberrantly stimulated alveolar/bronchiolar epithelial cells secrete the cytokines/growth factors responsible of the migration and proliferation of resident mesenchymal cells as well as of circulating progenitors (i.e., fibrocytes) and its transition to myofibroblasts. In this context, numerous studies have demonstrated that reactive alveolar/bronchiolar epithelial cells express platelet derived growth factor (PDGF), transforming growth factor beta-1 (TGF-β1), tumor necrosis factor alpha (TNF-α), endothelin-1, connective tissue growth factor (CTGF), and osteopontin all of them implicated in the development of pulmonary fibrosis. The alveolar epithelium may also generate a Th2-like pattern in IPF lungs, since it has been shown that these cells synthesize interleukin (IL)-4 while the expression of interferon-gamma (IFN-γ) is usually undetectable [3, 4, 11].

In addition, recent evidence also indicates that epithelial cells may contribute to the fibroblasts expansion through epithelial-mesenchymal transition (EMT). EMT represents a complex process by which polarized epithelial cells differentiate to contractile and motile mesenchymal cells, and plays a pivotal role in the generation of new tissue types during embryogenesis [12]. It has been demonstrated that after exposure to TGF-β1 in culture, alveolar epithelial cells undergo EMT as evidenced by loss of epithelial markers (aquaporin-5, E-cadherin) and upregulation of mesenchymal markers such as α-smooth muscle actin and type I collagen concomitant with transition to a fibroblast-like morphology [13]. Furthermore, two recent studies have demonstrated that EMT occurs *in vivo* in the lungs of patients with IPF [13, 14].

Nevertheless, the final common feature of any fibrotic lung disorder is the abnormal accumulation of extracellular matrix with the ultimate consequence of an extensive structural disorganization in the lung microenvironment where alveolar-capillary units are lost and replaced by scarring, bronchiolization of alveoli, and honeycombing. Importantly, many of the mechanisms behind this severe architectural remodeling involve an uncoordinated regulation and expression of a number of matrix metalloproteases which appear to be crucially implicated in the pathological processes from both, the inflammatory and the epithelial pathways.

Matrix metalloproteases (MMPs)

MMPs are the M10 family of endopeptidases that belong to the MA zinc-containing clan of metallopeptidases and to the metzincin subclan of proteases [15, 16].

MMPs also called matrixins consist of 25 members in mice and rats, and 23 enzymes in humans where are codified in 24 genes including duplicated MMP-23 genes [17]. Most of the MMPs are secreted enzymes, although there are some membrane types MMPs (MT-MMPs). Importantly however, some of the secreted MMPs have been found inside the cell and probably acting on intracellular substrates [18–21].

According to their structural and functional characteristics, MMPs family members have been classified into six different but closely related subgroups with fairly characteristic but often overlapping substrate specificities. This classification considers collagenases, gelatinases, stromelysins, matrilysins, membrane-type MMPs (MT-MMPs), and other MMPs [22]. On the other hand, another classification system has been proposed, based on MMP structure rather than on their substrate specificity, and include archetypal MMPs, matrilysins, gelatinases, and furin activable MMPs [23].

MMPs levels are usually low in normal adult tissues, and with some exceptions, their production and activity are maintained at practically undetectable levels. Thus, MMPs are tightly regulated at the transcriptional and post-transcriptional levels and their expression become elevated when there is a challenge to the system, such as wound healing, repair or remodeling processes, in diseased tissues, and even in several cell types grown in culture [24].

MMPs are also regulated by activation of the precursor zymogens and inhibition by endogenous inhibitors, tissue inhibitors of metalloproteases (TIMPs). The extracellular activation of most MMPs is regulated by a proteolytic cascade and can be initiated by other already activated MMPs or by several serine proteases. The TIMPs are the main natural inhibitors of MMPs and ADAMs (ADAM: a disintegrin and metallprotease) and, as such, they tightly control metalloproteases catalytic activity. Thus, the balance between MMPs and TIMPs are critical for the eventual ECM remodeling. However, although MMPs inhibition is a major physiological function of TIMPs, strong evidence indicates that TIMPs are multifunctional proteins that also evolve to fill non-inhibitory functions.

The contribution of MMPs to extracellular matrix remodeling is complex. It is well known that the MMPs are not only responsible of matrix degradation but a large body of evidence sustain that this activity is only a part of their biological function in tissues. MMPs have the ability to process an increasing list of bioactive mediators such as growth factors, cytokines, chemokines, and cell-surface-receptors modulating their activity either by direct cleavage, or releasing them from extracellular matrix bound stores. Clearly, their multifunctional effects increase the complexity of molecular actions of MMPs in the regulation of diverse cellular processes.

Matrix metalloproteases in fibrotic lung remodeling

Several MMPs have been shown to participate in the aberrant extracellular matrix remodeling of pulmonary fibrosis [25].

Importantly, among the multiple proteins that can be potentially targeted by MMPs some are profibrogenic while others appear to have anti-fibrogenic activity. In this context, MMPs play an essential but complex role in several interrelated processes that take place in the pathogenesis of IPF including extracellular matrix remodeling, basement membrane disruption, epithelial apoptosis, cell migration, and angiogenesis.

The putative role of MMPs in pulmonary fibrosis has been presumed on one hand by the observations related with the overexpression of some of them in the human fibrotic lungs and on the other hand, by evidence revealed in animal models and by the use of transgenic and knockout mice.

The most common experimental model for lung fibrosis is the bleomycin induced lung injury in rodents. However, it is important to have in mind that this model represents a reversible process of the inflammatory pathway of lung fibrosis. In this context, it provides an insight for the mechanisms implicated in the inflammatory pathway, i.e., inflammation, fibrosis and resolution including remodeling and restitution of lung architecture. By contrast, at least IPF as discussed earlier is an irreversible process in which inflammation does not seem to play a pathogenic role. Thus, it is important to be cautious in the interpretation of the results in the animal model and its implications in the human disease.

The global expression pattern of MMPs in IPF using oligonucleotide microarrays has shown that MMP-1, MMP-2, MMP-7, MMP-9 and MMP-28 are significantly increased in IPF compared with controls [16, 26].

MMP-1 is increased despite the progressive deposit of interstitial collagens
MMP-1 also known as collagenase-1, fibroblast collagenase, and interstitial collagenase is the archetype of MMPs capable of degrading fibrillar collagens (types I and III). For that reason, the finding of an overexpression of interstitial collagenase

in IPF lungs, where the main characteristic is the excessive accumulation of fibrilar collagens might be considered as a paradox. Moreover, this enzyme has been implicated in diseases that, in contrast to fibrosis, are characterized by exaggerated extracellular matrix degradation, such as rheumatoid arthritis and lung emphysema [27, 28]. However, an important fact is that the localization of the enzyme has been observed mainly in reactive alveolar epithelial cells as well as bronchiolar epithelial cells lining honeycomb cystic spaces while it appears to be absent in the interstitial compartment, and in the fibroblastic foci, that are the areas of ongoing fibrogenesis [29, 30].

Interestingly, TGF-β and osteopontin, two potent fibrogenic mediators that are strongly expressed in alveolar epithelial cells in IPF lungs, have shown to induce downregulation of MMP-1 expression in lung fibroblasts *in vitro* [31, 32].

As MMP-1 does not have a precise ortholog in the adult mouse or rat, the possible pathological role of this enzyme cannot been explored in experimental models. However, no significant changes in MMP-13, the prevalent collagenase expressed in these rodents, have been found in bleomycin and silica induced lung fibrosis [33–34]. By contrast, rat lungs evolving to fibrosis by the effect of paraquat plus hyperoxia showed a significant reduction in the mRNA expression of both collagenases MMP-8 and MMP-13 [35].

The role of lung epithelial MMP-1 in IPF is presently unknown. Considering that the transgenic mice expressing human MMP-1 in alveolar epithelial cells develop emphysematous lesions [36], it is possible to speculate that this enzyme may contribute to the formation of the cystic spaces (honeycombing), characteristic of IPF. Microscopic honeycombing consists of dense fibrosis and complete loss of lung architecture that is replaced by cysts of varying size, and the cause of such devastating destruction is unknown.

On the other hand, MMP-1 has been shown to play a pivotal role in epithelial cell migration in skin wound healing [37]. The strong epithelial expression of MMP-1 in IPF lungs suggests that a similar process to wound healing might be occurring in this disease, although re-epithelialization seems to be unsuccessful at least partially because it appears to occur when alveoli structures have already disappeared.

MMP-7: a profibrotic piece in the puzzle
A similar role in epithelial cell migration may be attributed to matrilysin (MMP-7). MMP-7 is one of the genes most consistently elevated in fibrotic lungs [26, 32, 38, 39]. Interestingly, MMP-7 null mice are relatively protected from bleomycin induced fibrosis suggesting that matrilysin is a critical player of the tissue response in fibrosis [26].

MMP-7 lacks the hemopexin domain which has been related to its broad substrate affinity for extracellular matrix molecules such as basement membrane collagen type IV, aggrecan, laminin, fibronectin, gelatin, entactin, decorin, tenascin, vitronectin, osteonectin, elastin and osteopontin. Additionally, this enzyme processes numerous bioactive substrates including FASL, β4 integrin, E-cadherin, pro-HB-EGF, plasminogen, pro-TNF-α, syndecan, and IGFBP-3.

In IPF lungs, the increased immunoreactive protein is expressed primarily by the abnormal alveolar epithelium and the active protein was demonstrated by tissue zymography in IPF lungs [26]. It is important to consider that the expression pattern of a protease suggest potential functions when finding co-expression with candidate substrates. Thus, the colocalization of MMP-7 with osteopontin in alveolar epithelial cells from IPF lungs, as well as the significant interaction between both molecules suggests that MMP-7 and osteopontin interact to affect the IPF phenotype [32]. This hypothesis is supported by the findings that MMP-7 is induced and activated by osteopontin while the later is cleaved and activated by MMP-7 [33, 40].

The role of MMP-7 in lung fibrosis might be multiple considering its broad substrate specificity. Diverse studies support a role for this enzyme in apoptosis, inflammation and innate immunity, among others. It has also been proposed that shedding of E-cadherin ectodomain in bleomycin-injured lung epithelium, plays an important role in this animal model being critical for epithelial repair [41].

The gelatinases MMP-2 and MMP-9: pro or antifibrotic players?

Probably, MMP-2 and MMP-9 are the MMPs more analyzed in different human diseases. This situation derives from the easiness of the methodological approach for gelatinases activity through gelatin zymography. In this context, numerous evidence indicates that gelatinases (MMP-2 and MMP-9) participate in human interstitial lung diseases and in experimental models of lung fibrosis. Both gelatinases contain a fibronectin type II-like repeats within their catalytic domain, resulting in a higher binding affinity to gelatin and elastin.

MMP-2 (gelatinase A) degrades an extensive range of matrix and non-matrix substrates. It is effective primarily against type IV collagen and other basement membrane components although a weak ability to degrade stromal collagens has been also reported [25, 42]. MMP-2 has been proposed to have mainly anti-inflammatory and homeostatic functions, presumably by inactivating inflammatory chemokines [43] and by regulating connective tissue turnover [44].

MMP-2 gene expression has been found strongly upregulated in IPF tissues and the enzyme activity is usually increased in bronchoalveolar lavage fluids [30]. In addition, MMP-2 is frequently upregulated in experimental models of lung fibrosis and its overexpression as well as that of MMP-9 has been suspected to be implicated in basement membrane disruption [35, 42]. Increased MMP-2 protein has been consistently found in alveolar and basal bronchiolar epithelial cells as well as in the mesenchymal cells forming the fibroblastic foci [29, 30, 45, 46].

It has been well documented that the major physiological activators of proMMP-2 are members of the MT-MMP subfamily. Therefore, the increased activated form of MMP-2 usually found in bronchoalveolar lavage of IPF patients might be attributed to the action of these enzymes. Interestingly some MT-MMPs localize in IPF lungs in similar sites than MMP-2. Thus, MT1- and MT2-MMPs are expressed by alveolar epithelial cells, MT3-MMP by fibroblasts from fibroblastic foci as well as alveolar epithelial cells, and MT5-MMP by basal bronchiolar epithelial cells and in areas of squamous metaplasia [46].

The main structural difference between MMP-9 and MMP-2 is the presence of an extensively O-glycosylated (OG) domain in MMP-9 [47]. MMP-9 gene expression and protein have been also found elevated in human and experimental lung fibrosis [26, 29, 30, 45, 48–50] and interestingly enough an increase in the active form in BAL has been associated to an accelerated clinical phenotype [51].

The disruption of basement membranes by MMP-2 and MMP-9 may contribute to the aberrant lung remodeling by at least two mechanisms, preventing alveolar re-epithelialization, and enhancing the fibroblast invasion into the alveolar spaces [52]. However, studies in experimental models show a more complex picture. Thus, after bleomycin installation, MMP-9 null mice develop a similar fibrotic response to that observed in the wild-type littermates. However, the lungs of MMP-9 deficient mice showed minimal alveolar bronchiolization suggesting that gelatinase B facilitates migration of Clara cells and other bronchiolar cells into the regions of alveolar injury [53]. On the other hand, a recent study in transgenic mice overexpressing MMP-9 by alveolar macrophages showed that transgenic mice develop a less intense lung fibrotic reaction after bleomycin instillation [54]. The reduction of profibrotic mediators such as TIMP-1 and IGFBP-3 in the MMP-9 transgenic mice was identified as potential mechanisms of the diminished fibrotic response. These findings highlight the divergent roles that this enzyme may have either promoting or attenuating the fibrotic response. Interestingly, a recent study also, demonstrated a protective role of MMP-9 but not of MMP-7, in O(3)-induced lung neutrophilic inflammation and hyperpermeability [55]. Supporting this point of view, different studies indicate that MMP-9 may play a dual role in the migration of inflammatory cells to sites of injury. For example, regarding neutrophil infiltration, it has been shown that MMP-9 processes IL-8, a potent human neutrophil chemokine, into a more active mediator and in mouse cleaves GCP-2/LIX, a potent mouse CXC chemokine also potentiating its biological activity [56, 57]. On the other hand however, MMP-9 also degrades neutrophilic chemokines like GROα, CTAP-III, and PF-4 [56, 57].

Importantly, several studies have demonstrated the relevance of an early neutrophilic response in the development of lung fibrosis. Less neutrophils in the BAL has been associated with less lung fibrosis in gamma-glutamyl transpeptidase deficient mice and in MMP-7 deficient mice as well as in transgenic mice overexpressing MMP-9 [41, 50, 54]. However, a recent report revealed that early depletion of neutrophils with anti-mouse PMN antibody and the associated modifications in pro-MMP-9/TIMP-1 lung balance did not alter susceptibility to bleomycin-induced pulmonary fibrosis [58].

In this context, it is still difficult to have a clear picture regarding MMP-9 role in the pathogenesis of pulmonary fibrosis since its expression in different type of cells (neutrophils *versus* macrophages *versus* epithelial cells), the location of the activity (alveolar septum *versus* basement membrane), the target substrate(s) and the level of expression/activity of MMP-9 (too little or too much) may turn on different pathways in response to injury.

Tissue inhibitors of metalloproteinases in pulmonary fibrosis

A major role of TIMPs in tissue remodeling is their ability to inhibit MMPs. Therefore, changes in their production and localization may contribute to modifications in the enzymatic activity of MMPs in the lung microenvironment.

The growing evidence of the multiple diverse roles of MMPs in various cellular processes is accompanied with the complexity of the functions of TIMPs under physiological and pathological conditions. Thus, in addition to their regulation of MMP-mediated degradation, TIMPs present a variety of MMP-independent actions and participate in the regulation of cell death, cell proliferation, and angiogenesis among others [59].

The TIMP family consists of four distinct small secreted proteins (TIMP-1, 2, 3, and 4) of about 20–29 kDa, that reversibly inhibit the MMPs in a 1:1 stoichiometric fashion [60, 61].

The balance between MMPs and TIMPs has been implicated in both normal and pathological events including wound healing and tissue remodeling. However, studies regarding the role of TIMPs in human lung fibrosis including IPF are scanty. Global gene expression has revealed that while some matrix metalloproteinases are among the most highly expressed genes in IPF, TIMPs did not appear among the most elevated genes [26]. This result is surprising since the opposite situation would be theoretically expected given that TIMPs overexpression favors local extracellular matrix accumulation. Nevertheless, although not as high as MMPs, TIMP-1 and TIMP-3 were significantly increased in IPF lungs as compared to controls [62].

Regarding the location, a different picture emerges from the TIMPs studies than the mentioned for interstitial collagenases. Thus, noticeable interstitial presence of TIMPs 1, 2, 3 and 4 compared with collagenases was found *in vivo* in human lung fibrosis, suggesting that a non-degrading fibrillar collagen microenvironment is present, at least in the interstitial compartment, in IPF [30] (Fig. 1).

TIMP-1 appears expressed in the stromal matrix and fibroblasts as well as in areas of epithelial cells in human lung fibrosis [30]. Interestingly, a recent study identified CD63, a member of the tetraspanin family, as a TIMP-1 interacting protein, and confocal microscopic analysis confirmed CD63 interactions with TIMP-1, integrin β1, and their co-localizations on the cell surface of human breast epithelial MCF10A cells. shRNA-mediated CD63 downregulation effectively reduced TIMP-1

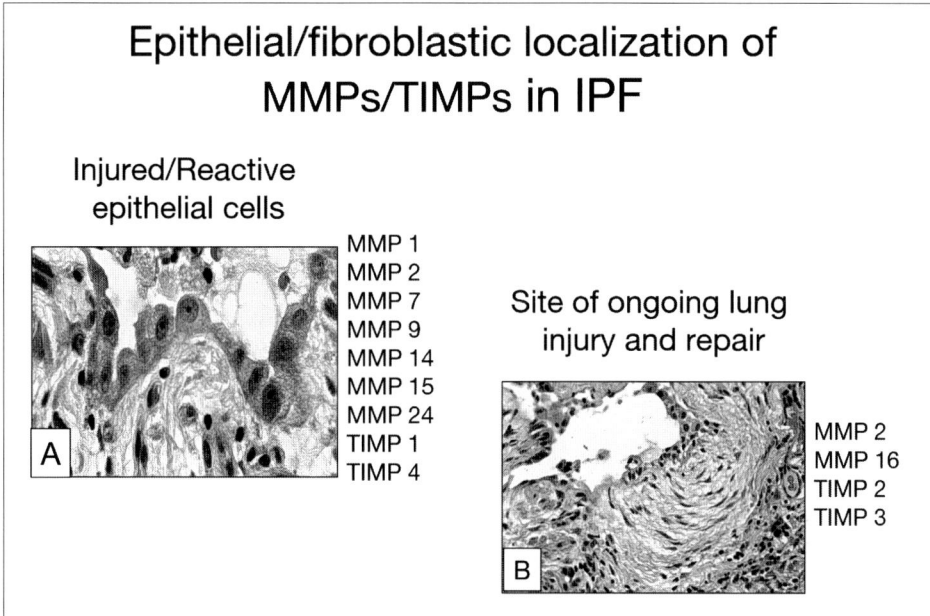

Figure 1
Aberrant alveolar epithelial cell phenotypes (panel A) and the formation of fibroblastic/ myofibroblastic foci (panel B) are characteristic changes in IPF lungs. Importantly, a different cluster of matrix metalloproteinases and tissue inhibitors are expressed by these cell-types during the development of the disease.

binding to the cell surface, and consequently reversed TIMP-1-mediated integrin $\beta1$ activation, cell survival signaling and apoptosis inhibition [63].

Different studies in animal models of lung fibrosis, have strongly suggested that TIMP-1 plays an important profibrotic role. For example, transient exposure to active TGF-β1 induced a severe lung fibrosis in C57BL/6 mice compared with Balb/c mice which was primarily due to a noteworthy upregulation of TIMP-1 in the lungs of the susceptible mice [64]. Similar results were obtained with the same strain of mice exposed to bleomycin and in rats injured with paraquat plus hyperoxia [35, 65]. Likewise, mice deficient in either TNF-α receptor p55 or p75 were protected against silica-induced lung fibrosis which was associated with a decreased expression of TIMP-1 [34].

Surprisingly however, TIMP-1 deficiency did not modify collagen accumulation after bleomycin-induced lung injury [66]. Similar results were obtained in a recent study where C57BL/6 mice overexpressed TIMP-1 in their lungs *via* SPC promoter [67]. In this C57BL/6 transgenic mouse, epithelial expression of TIMP-1 did not alter sensitivity to bleomycin and similar levels of hydroxyproline as their wild type counterparts were observed in the lung tissue [68]. On the other hand, interestingly, TIMP-1-deficient mice showed an exaggerated gelatinase-B activity in the alveolar compartment, and in a similar situation transgenic mice overexpressing MMP-9 displayed a diminished TIMP-1 protein expression. These results indicate a strong interrelation between MMP-9 and TIMP-1. It has been shown that the C-terminal domain of MMP-9, also known as the hemopexin domain, is an exosite for binding of TIMP-1 [67].

Of particular interest is the evidence that TIMP-2 is primarily expressed by subepithelial fibroblast/myofibroblast foci in IPF lungs [29, 30]. It is well known that these distinct clusters of mesenchymal cells represent areas of active fibrogenesis and appear to play a crucial role in the pathogenesis of IPF [4]. The expression of TIMP-2 in these peculiar clusters may have several profibrotic effects including collagenase activity inhibition, stimulation of fibroblast proliferation, and activation of latent gelatinase A. Actually, TIMP-2 has been co-localized with nuclear markers of cell proliferation suggesting a role in the expansion of the fibroblast/myofibroblast cell population [30]. This process may explain, at least partially, the survival of mesenchymal cell populations in the fibroblast foci, against the expected cell death as it is observed in a normal wound healing model [69]. Supporting a likely deleterious role of TIMP-1 and 2 in lung fibrosis is the finding that rats showing spontaneous recovery from experimental cirrhosis exhibit a rapid decrease of both inhibitors together with an increase of collagenolytic activity, and apoptosis of hepatic stellate cells [70]. Likewise, in human liver fibrosis, serum levels of TIMP-1 and 2 are related to the histological degrees of hepatic fibrosis [71, 72].

The expression of TIMP-3 has been shown to be increased in IPF lungs. This inhibitor was revealed strongly staining the elastic lamina of vessels and also localized in fibroblasts from fibroblastic foci as well as in endothelial cells [30, 62]. The expression of TIMP-3 in the fibroblastic foci suggests that this inhibitor may be also implicated in the low degradative microenvironment that seems to characterize these foci. Supporting a profibrotic role of TIMP-3, it has been shown that the homozygous Timp-3-null animals develop spontaneous air space enlargement in the lung and lungs from aged null animals have reduced abundance of collagen and enhanced degradation of collagen [73].

Interestingly, it was recently reported that TIMP-3 expression is approximately two-fold higher in bleomycin-induced pulmonary fibrosis resistant BALB/c mice than in sensitive C57BL/6 mice [74]. As TIMP-3 is the major physiological inhibitor of tumor necrosis factor converting enzyme (TACE) and a crucial innate negative regulator of TNF-α, it was suggested that the lower expression of TIMP-3 observed in C57BL/6 mice could be associated with a TNF-α-mediated inflammatory syndrome.

Additionally, TIMP-3 is a potent inhibitor of angiogenesis that blocks the binding of VEGF to VEGF receptor-2, property that seems to be independent of its

MMP-inhibitory activity [75]. Importantly, angiogenesis appears to be strongly inhibited in the fibroblastic foci, suggesting that this process can be relevant for fibrogenesis [76]. Thus, the expression of the angiostatic TIMP-3 by fibroblasts from these fibroblastic foci might play a role in the absence of capillaries that seems to characterize these foci.

The molecular mechanisms involved in the increased expression of the TIMPs are not completely elucidated but it is well known that TGF-β1, a pivotal profibrotic mediator causes a strong upregulation of TIMP-1 and TIMP-3. Induction of TIMP-3 in lung fibroblasts, is blocked by genistein, and partially by SB203580, indicating that TGF-β-induced TIMP-3 is mediated through phosphorylation on tyrosine residues [62].

Concluding remarks

Lung fibrosis is characterized by the expansion of the fibroblasts/myofibroblasts population in the interalveolar septa and alveolar spaces together with an aberrant increase of extracellular matrix. The thickening of the septa and occupation of the alveolar lumen results in the loss of lung compliance and decreased gas exchange. Our knowledge about the pathogenic mechanisms and pathways involved in the development of lung fibrosis is improving, but many intriguing questions remain unsolved. Matrix metalloproteinases are involved in the remodeling of several components of the extracellular matrix including basement membranes, and play a pivotal role in the fibrotic response of the lungs. However, the action of the MMPs is not restricted to the extracellular matrices and a forceful body of evidence has revealed that they also cleave numerous non-ECM molecules, including chemokines, cytokines, and growth factors, either activating or inactivating them. These multifunctional effects and versatility make a difficult task to find out the biological consequences of the up- or downregulation of these enzymes (and their tissue inhibitors) in the lung microenvironment during the dynamics of the fibrotic response. Futures translational research will hopefully allow us to elucidate the complex molecular mechanisms implicated in the pathogenesis of lung fibrosis.

References

1   Ryu JH, Daniels CE, Hartman TE, Yi ES (2007) Diagnosis of interstitial lung diseases. *Mayo Clin Proc* 82: 976–986

2   Green FH (2002) Overview of pulmonary fibrosis. *Chest* 122(6 Suppl): 334S–339S

3   Pardo A, Selman M (2002) Molecular mechanisms of pulmonary fibrosis. *Front Biosci* 7: d1743–1761

4   Selman M, King TE, Pardo A (2001) Idiopathic pulmonary fibrosis: prevailing and

evolving hypotheses about its pathogenesis and implications for therapy. *Ann Intern Med* 134: 136–151

5   Luzina IG, Todd NW, Iacono AT, Atamas SP (2008) Roles of T lymphocytes in pulmonary fibrosis. *J Leukoc Biol* 83: 237–244

6   Stramer BM, Mori R, Martin P (2007) The inflammation-fibrosis link? A Jekyll and Hyde role for blood cells during wound repair. *J Invest Dermatol* 127: 1009–1017

7   Huaux F (2007) New developments in the understanding of immunology in silicosis. *Curr Opin Allergy Clin Immunol* 7: 168–173

8   Keane MP, Strieter RM, Lynch JP 3rd, Belperio JA (2006) Inflammation and angiogenesis in fibrotic lung disease. *Semin Respir Crit Care Med* 27: 589–599

9   Garrood T, Lee L, Pitzalis C (2006) Molecular mechanisms of cell recruitment to inflammatory sites: general and tissue-specific pathways. *Rheumatology (Oxford)* 45: 250–260

10   Rao RM, Shaw SK, Kim M, Luscinskas FW (2005) Emerging topics in the regulation of leukocyte transendothelial migration. *Microcirculation* 12: 83–89

11   Selman M, Pardo A (2006) Role of epithelial cells in idiopathic pulmonary fibrosis: from innocent targets to serial killers. *Proc Am Thorac Soc* 3: 364–372

12   Lee JM, Dedhar S, Kalluri R, Thompson EW (2006) The epithelial-mesenchymal transition: new insights in signaling, development, and disease. *J Cell Biol* 172: 973–981

13   Willis BC, Liebler JM, Luby-Phelps K, Nicholson AG, Crandall ED, du Bois RM, Borok Z (2005) Induction of epithelial-mesenchymal transition in alveolar epithelial cells by transforming growth factor-beta1: potential role in idiopathic pulmonary fibrosis. *Am J Pathol* 166: 1321–1332

14   Kim KK, Kugler MC, Wolters PJ, Robillard L, Galvez MG, Brumwell AN, Sheppard D, Chapman HA (2006) Alveolar epithelial cell mesenchymal transition develops *in vivo* during pulmonary fibrosis and is regulated by the extracellular matrix. *Proc Natl Acad Sci USA* 103: 13180–13185

15   Rawlings ND, Morton FR, Barrett AJ (2006) MEROPS: the peptidase database. *Nucleic Acids Res* 34: D270–D272

16   Pardo A, Selman M, Kaminski N (2008) Approaching the degradome in idiopathic pulmonary fibrosis. *Int J Biochem Cell Biol* 40: 1141–1155

17   Puente XS, Sanchez LM, Overall CM, Lopez-Otin C (2003) Human and mouse proteases: a comparative genomic approach. *Nat Rev Genet* 4: 544–558

18   Limb GA, Matter K, Murphy G, Cambrey AD, Bishop PN, Morris GE, Khaw PT (2005) Matrix metalloproteinase-1 associates with intracellular organelles and confers resistance to lamin A/C degradation during apoptosis. *Am J Pathol* 166: 1555–1563

19   Wang W, Schulze CJ, Suarez-Pinzon WL, Dyck JR, Sawicki G, Schulz R ((2002) Intracellular action of matrix metalloproteinase-2 accounts for acute myocardial ischemia and reperfusion injury. *Circulation* 106: 1543–1549

20   Kwan JA, Schulze CJ, Wang W, Leon H, Sariahmetoglu M, Sung M, Sawicka J, Sims DE, Sawicki G, Schulz R (2004) Matrix metalloproteinase-2 (MMP-2) is present in the

nucleus of cardiac myocytes and is capable of cleaving poly (ADP-ribose) polymerase (PARP) *in vitro*. *FASEB J* 18: 690–692

21 Luo D, Mari B, Stoll I, Anglard P (2002) Alternative splicing and promoter usage generates an intracellular stromelysin 3 isoform directly translated as an active matrix metalloproteinase. *J Biol Chem* 277: 25527–25536

22 Brinckerhoff CE, Matrisian LM (2002) Matrix metalloproteinases: a tail of a frog that became a prince. *Nat Rev Mol Cell Biol* 3: 207–214

23 Folgueras AR, Pendas AM, Sanchez LM, Lopez-Otin C (2004) Matrix metalloproteinases in cancer: from new functions to improved inhibition strategies. *Int J Dev Biol* 48: 411–424

24 Sternlicht MD, Werb Z (2001) How matrix metalloproteinases regulate cell behavior. *Annu Rev Cell Dev Biol* 17: 463–516

25 Pardo A, Selman M (2006) Matrix metalloproteases in aberrant fibrotic tissue remodeling. *Proc Am Thorac Soc* 3: 383–388

26 Zuo F, Kaminski N, Eugui E, Allard J, Yakhini Z, Ben-Dor A, Lollini L, Morris D, Kim Y, DeLustro B et al (2002) Gene expression analysis reveals matrilysin as a key regulator of pulmonary fibrosis in mice and humans. *Proc Natl Acad Sci USA* 999: 6292–6297

27 Segura-Valdez L, Pardo A, Gaxiola M, Uhal BD, Becerril C, Selman M (2000) Upregulation of gelatinases A and B, collagenases 1 and 2, and increased parenchymal cell death in COPD. *Chest* 117: 684–694

28 Vincenti MP, Brinckerhoff CE (2002) Transcriptional regulation of collagenase (MMP-1, MMP-13) genes in arthritis: integration of complex signaling pathways for the recruitment of gene-specific transcription factors. *Arthritis Res* 4: 157–164

29 Fukuda Y, Ishizaki M, Kudoh S, Kitaichi M, Yamanaka N (1998) Localization of matrix metalloproteinases-1, -2, and -9 and tissue inhibitor of metalloproteinase-2 in interstitial lung diseases. *Lab Invest* 78: 687–698

30 Selman M, Ruiz V, Cabrera S, Segura L, Ramírez R, Barrios R, Pardo A (2000) TIMP-1, -2, -3 and -4 in idiopathic pulmonary fibrosis. A prevailing non degradative lung microenvironment? *Am J Physiol* 279: L562–L574

31 Hall MC, Young DA, Waters JG, Rowan AD, Chantry A, Edwards DR, Clark IM (2003) The comparative role of activator protein 1 and Smad factors in the regulation of Timp-1 and MMP-1 gene expression by transforming growth factor-beta 1. *J Biol Chem* 278: 10304–10313

32 Pardo A, Gibson K, Cisneros J, Richards TJ, Yang Y, Becerril C, Yousem S, Herrera I, Ruiz V, Selman M et al (2005) Up-regulation and profibrotic role of osteopontin in human idiopathic pulmonary fibrosis. *PLoS Med* 2: e251

33 Swiderski RE, Dencoff JE, Floerchinger CS, Shapiro SD, Hunninghake GW (1998) Differential expression of extracellular matrix remodeling genes in a murine model of bleomycin-induced pulmonary fibrosis. *Am J Pathol* 152: 821–828

34 Ortiz LA, Lasky J, Gozal E, Ruiz V, Lungarella G, Cavarra E, Brody AR, Friedman M, Pardo A, Selman M (2001) Tumor necrosis factor receptor deficiency alters matrix

metalloproteinase 13/tissue inhibitor of metalloproteinase 1 expression in murine silicosis. *Am J Respir Crit Care Med* 163: 244–252

35 Ruiz V, Ordonez RM, Berumen J, Ramirez R, Uhal B, Becerril C, Pardo A, Selman M (2003) Unbalanced collagenase/TIMP-1 expression and epithelial apoptosis in experimental lung fibrosis. *Am J Physiol Lung Cell Mol Physiol* 285: L1026–1036

36 D'Armiento J, Dalal SS, Okada Y, Berg RA, Chada K (1992) Collagenase expression in the lungs of transgenic mice causes pulmonary emphysema. *Cell* 71: 955–961

37 Pilcher BK, Dumin JA, Sudbeck BD, Krane SM, Welgus HG, Parks WC (1997) The activity of collagenase-1 is required for keratinocyte migration on a type I collagen matrix. *J Cell Biol* 137: 1445–1457

38 Selman M, Pardo A, Barrera L, Estrada A, Watson SR, Wilson K, Aziz N, Kaminski N, Zlotnik A (2006) Gene expression profiles distinguish idiopathic pulmonary fibrosis from hypersensitivity pneumonitis. *Am J Respir Crit Care Med* 173: 188–198

39 Vuorinen K, Myllarniemi M, Lammi L, Piirila P, Rytila P, Salmenkivi K, Kinnula VL (2007) Elevated matrilysin levels in bronchoalveolar lavage fluid do not distinguish idiopathic pulmonary fibrosis from other interstitial lung diseases. *APMIS* 115, 969–975

40 Agnihotri R, Crawford HC, Haro H, Matrisian LM, Havrda MC, Liaw L (2001) Osteopontin, a novel substrate for matrix metalloproteinase-3 (stromelysin-1) and matrix metalloproteinase-7 (matrilysin). *J Biol Chem* 276: 28261–28267

41 Li Q, Park PW, Wilson CL, Parks WC (2002) Matrilysin shedding of syndecan-1 regulates chemokine mobilization and transepithelial efflux of neutrophils in acute lung injury. *Cell* 111: 635–646

42 Patterson ML, Atkinson SJ, Knäuper V, Murphy G (2001) Specific collagenolysis by gelatinase A, MMP-2, is determined by the hemopexin domain and not the fibronectin-like domain. *FEBS Lett* 503: 158–162

43 McQuibban GA, Gong JH, Tam EM, McCulloch CA, Clark-Lewis I, Overall CM (2000) Inflammation dampened by gelatinase A cleavage of monocyte chemoattractant protein-3. *Science* 289: 1202–1206

44 Monaco S, Sparano V, Gioia M, Sbardella D, Di Pierro D, Marini S, Coletta M (2006) Enzymatic processing of collagen IV by MMP-2 (gelatinase A) affects neutrophil migration and it is modulated by extracatalytic domains. *Protein Sci* 15: 2805–2815

45 Hayashi T, Stetler-Stevenson WG, Fleming MV, Fishback N, Koss MN, Liotta LA, Ferrans VJ, Travis WD (1996) Immunohistochemical study of metalloproteinases and their tissue inhibitors in the lungs of patients with diffuse alveolar damage and idiopathic pulmonary fibrosis. *Am J Pathol* 149: 1241–1256

46 García-Alvarez J, Ramirez R, Sampieri CL, Nuttall RK, Edwards DR, Selman M, Pardo A (2006) Membrane type-matrix metalloproteinases in idiopathic pulmonary fibrosis. *Sarcoidosis Vasc Diffuse Lung Dis* 23: 13–21

47 Van den Steen PE, Van Aelst I, Hvidberg V, Piccard H, Fiten P, Jacobsen C, Moestrup SK, Fry S, Royle L, Wormald MR et al (2006) The hemopexin and O-glycosylated domains tune gelatinase B/MMP-9 bioavailability *via* inhibition and binding to cargo receptors. *J Biol Chem* 281: 18626–18637

48   Pérez-Ramos J, Segura L, Ramírez R, Vanda B, Selman M, Pardo A (1999) Matrix metalloproteinases 2, 9, and 13 and tissue inhibitor of metalloproteinases 1 and 2 in early and late lesions of experimental lung silicosis. *Am J Respir Crit Care Med* 160: 1274–1282

49   Cisneros-Lira J, Gaxiola M, Ramos C, Selman M, Pardo A (2003) Cigarette smoke exposure potentiates bleomycin-induced lung fibrosis in guinea pigs. *Am J Physiol Lung Cell Mol Physiol* 285: L949–956

50   Pardo A, Ruiz V, Arreola JL, Ramírez R, Cisneros-Lira J, Gaxiola M, Barrios R, Kala SV, Lieberman MW, Selman M (2003) Bleomycin-induced pulmonary fibrosis is attenuated in g_glutamyl transpeptidase-deficient mice. *Am J Respir Crit Care Med* 167: 925–932

51   Selman M, Carrillo G, Estrada A, Mejia M, Becerril C, Cisneros J, Gaxiola M, Pérez-Padilla R, Navarro C, Richards T et al (2207) Accelerated variant of idiopathic pulmonary fibrosis: clinical behavior and gene expression pattern. *PLoS ONE* 2: e482

52   Atkinson JJ, Senior RM (2003) Matrix metalloproteinase-9 in lung remodeling. *Am J Respir Cell Mol Biol* 28: 12–24

53   Betsuyaku T, Fukuda Y, Parks WC, Shipley JM, Senior RM (2000) Gelatinase B is required for alveolar bronchiolization after intratracheal bleomycin. *Am J Pathol* 157: 525–535

54   Cabrera S, Gaxiola M, Arreola JL, Ramírez R, Jara P, D'Armiento J, Selman M, Pardo A (2007) Overexpression of MMP9 in macrophages attenuates pulmonary fibrosis induced by bleomycin. *Int J Biochem Cell Biol* 39: 2324–2338

55   Yoon HK, Cho HY, Kleeberger SR (2007) Protective role of matrix metalloproteinase-9 in ozone-induced airway inflammation. *Environ Health Perspect* 115: 1557–1563

56   Van den Steen PE, Proost P, Wuyts A, Van Damme J, Opdenakker G (2000) Neutrophil gelatinase B potentiates interleukin-8 tenfold by aminoterminal processing, whereas it degrades CTAP-III, PF-4, and GRO- and leaves RANTES and MCP-2 intact. *Blood* 96: 2673–2681

57   Van Den Steen PE, Wuyts A, Husson SJ, Proost P, Van Damme J, Opdenakker G (2003) Gelatinase B/MMP-9 and neutrophil collagenase/MMP-8 process the chemokines human GCP-2/CXCL6, ENA-78/CXCL5 and mouse GCP-2/LIX and modulate their physiological activities. *Eur J Biochem* 270: 3739–3749

58   Manoury B, Nénan S, Guénon I, Lagente V, Boichot E (2007) Influence of early neutrophil depletion on MMPs/TIMP-1 balance in bleomycin-induced lung fibrosis. *Int Immunopharmacol* 7: 900–911

59   Chirco R, Liu XW, Jung KK, Kim HR (2006) Novel functions of TIMPs in cell signaling. *Cancer Metastasis Rev* 25: 99–113

60   Gomez DE, Alonso DF, Yoshiji H, Thorgeirsson UP (1997) Tissue inhibitors of metalloproteinases: structure, regulation and biological functions. *Eur J Cell Biol* 74: 111–122

61   Murphy G, Willenbrock F (1995) Tissue inhibitors of matrix metalloendopeptidases. *Methods Enzymol* 248: 496–510

62   García-Alvarez J, Ramirez R, Checa M, Nuttall RK, Sampieri CL, Edwards DR, Selman M, Pardo A (2006) Tissue inhibitor of metalloproteinase-3 is up-regulated by transform-

ing growth factor-beta1 *in vitro* and expressed in fibroblastic foci *in vivo* in idiopathic pulmonary fibrosis. *Exp Lung Res* 32: 201–214

63 Jung KK, Liu XW, Chirco R, Fridman R, Kim HR (2006) Identification of CD63 as a tissue inhibitor of metalloproteinase-1 interacting cell surface protein. *EMBO J* 25: 3934–3942

64 Kolb M, Bonniaud P, Galt T, Sime PJ, Kelly MM, Margetts PJ, Gauldie J (2002) Differences in the fibrogenic response after transfer of active transforming growth factor-beta1 gene to lungs of 'fibrosisprone' and 'fibrosis-resistant' mouse strains. *Am J Respir Cell Mol Biol* 27: 141–150

65 Manoury B, Caulet-Maugendre S, Guénon I, Lagente V, Boichot E (2006) TIMP-1 is a key factor of fibrogenic response to bleomycin in mouse lung. *Int J Immunopathol Pharmacol* 19: 471–487

66 Kim KH, Burkhart K, Chen P, Frevert CW, Randolph-Habecker J, Hackman RC, Soloway PD, Madtes DK (2005) Tissue inhibitor of metalloproteinase-1 deficiency amplifies acute lung injury in bleomycin-exposed mice. *Am J Respir Cell Mol Biol* 33: 271–279

67 Van den Steen PE, Van Aelst I, Hvidberg V, Piccard H, Fiten P, Jacobsen C, Moestrup SK, Fry S, Royle L, Wormald MR et al (2006) The hemopexin and O-glycosylated domains tune gelatinase B/MMP-9 bioavailability *via* inhibition and binding to cargo receptors. *J Biol Chem* 281: 18626–18637

68 Fattman CL, Gambelli F, Hoyle GW, Pitt BR, Ortiz LA (2008) Epithelial expression of TIMP1 does not alter sensitivity to bleomycin-induced lung injury in C57BL/6 mice. *Am J Physiol Lung Cell Mol Physiol* 294: L572–581

69 Desmouliere A, Redard M, Darby I, Gabianni G (1995) Apoptosis mediates the decrease in cellularity during the transition between granulation tissue and scar. *Am J Pathol* 146: 56–66

70 Iredale JP, Benyon RC, Pickering J, McCullen M, Northrop M, Pawley S, Hovell C, Arthur MJ (1998) Mechanisms of spontaneous resolution of rat liver fibrosis. Hepatic stellate cell apoptosis and reduced hepatic expression of metalloproteinase inhibitors. *J Clin Invest* 102: 538–549

71 Murawaki Y, Ikuta Y, Kawasaki H (1999) Clinical usefulness of serum tissue inhibitor of metalloproteinases (TIMP)-2 assay in patients with chronic liver disease in comparison with serum TIMP-1. *Clin Chim Acta* 281: 109–120

72 Hemmann S, Graf J, Roderfeld M, Roeb E (2007) Expression of MMPs and TIMPs in liver fibrosis – a systematic review with special emphasis on anti-fibrotic strategies. *J Hepatol* 46: 955–975

73 Leco KJ, Waterhouse P, Sanchez OH, Gowing KL, Poole AR, Wakeham A, Mak TW, Khokha R (2001) Spontaneous air space enlargement in the lungs of mice lacking tissue inhibitor of metalloproteinases-3 (TIMP-3). *J Clin Invest* 108: 817–829

74 Pottier N, Chupin C, Defamie V, Cardinaud B, Sutherland R, Rios G, Gauthier F, Wolters PJ, Berthiaume Y, Barbry P et al (2007) Relationships between early inflammatory response to bleomycin and sensitivity to lung fibrosis: a role for dipeptidyl-peptidase

I and tissue inhibitor of metalloproteinase-3? *Am J Respir Crit Care Med* 176: 1098–1107

75 Qi JH, Ebrahem Q, Moore N, Murphy G, Claesson-Welsh L, Bond M, Baker A, Anand-Apte B (2003) A novel function for tissue inhibitor of metalloproteinases-3 (TIMP3): inhibition of angiogenesis by blockage of VEGF binding to VEGF receptor-2. *Nat Med* 9: 407–415

76 Cosgrove GP, Brown KK, Schiemann WP, Serls AE, Parr JE, Geraci MW, Schwarz MI, Cool CD, Worthen GS (2004) Pigment epithelium-derived factor in idiopathic pulmonary fibrosis: a role in aberrant angiogenesis. *Am J Respir Crit Care Med* 170: 242–251

Anti-inflammatory properties of MMP inhibitors in experimental models of chronic obstructive pulmonary disease and lung inflammation

Catherine le Quément[1], Vincent Lagente[1], Isabelle Guénon[1], Valeria Muzio[2], Jean-Yves Gillon[3] and Elisabeth Boichot[1]

[1]INSERM U620, Faculté de Pharmacie, Université de Rennes 1, 2, avenue du Pr. Léon Bernard, 35043 Rennes, France; [2]LCG-RBM, Istituto di Ricerche Biomediche "Antoine Marxer", 1 Via Ribes, 10010 Colleretto Giacosa, Italy; [3]Merck-Serono International S.A., 9 Chemin des Mines, 1211 Geneva 20, Switzerland

Abstract

Matrix metalloproteinases (MMPs) are a group of proteases known to regulate the turnover of extracellular matrix and thus are suggested to be important in the process of lung disease associated with tissue remodelling. Furthermore, the concept that modulation of airway remodelling including excessive proteolysis damage of the tissue, may be of interest as a basis for future treatment. Among the metalloproteinases (MMPs) family, macrophage elastase (MMP-12) is able to degrade extracellular matrix components such as elastin and is involved in tissue remodelling processes in chronic obstructive pulmonary disease including emphysema. Recent studies using broad spectrum MMP or dual MMP-9/MMP-12 inhibitors have demonstrated a reduction in both inflammatory process and airspace enlargement in lung tissue. In the present chapter, we also report the inhibitory activity of a new MMP-9/MMP-12 inhibitor, AS112108, on acute lung inflammatory processes induced by cigarette smoke.

Introduction

Chronic obstructive pulmonary disease (COPD) is one of the major causes of mortality and morbidity in the developed countries and its prevalence is still increasing [1]. The major triggering factor is cigarette smoking, which accounts for 80–90% of the COPD cases. However, in the population of smokers, only 15% of the subjects develop chronic airflow limitation [2].

COPD is characterised by the presence of a partially reversible airflow obstruction. This pathology is also associated with an airway inflammatory process characterised by an accumulation of inflammatory cells such as macrophages and

Matrix Metalloproteinases in Tissue Remodelling and Inflammation, edited by Vincent Lagente and Elisabeth Boichot neutrophils. Indeed, it has been shown that cigarette smoke consistently produces an increase in the neutrophil number in bronchoalveolar lavage fluid and in lung tissue [3, 4]. Macrophage numbers are also elevated in the lungs of smokers and of patients with COPD where they accumulate in the alveoli, bronchioli and small airways. Furthermore, there is a positive correlation between macrophage number in the alveolar walls and the mild-to-moderate emphysema status in patients with COPD [5]. It is generally believed that the development of emphysema reflects a relative excess of cell-derived proteases that degrade the connective tissue of the lung and a relative paucity of antiproteolytic defenses. This theory is often referred to as the 'protease–antiprotease imbalance' hypothesis and involves mainly serine proteases like neutrophil elastase and matrix metalloproteinases (MMPs).

Matrix metalloproteinases (MMPs), also called matrixins, are zinc-dependent endopeptidases, known for their ability to cleave one or several constituents of the extracellular matrix. They represent a large family of proteases that share common structural and functional elements and are produced from different genes. These enzymes are primarily distinguished from other classes of proteinases by their dependence on metal ions and neutral pH for activity. Zymogen forms of the MMPs (pro-MMPs) are secreted into the extracellular space from a large number of cell types, where activation of the pro-MMPs in the local microenvironment can result in discrete alterations in the tissue architecture. There is also increasing evidence for a role of MMPs in COPD, especially of MMP-9 and MMP-12 [6, 7].

Involvement of MMP-9 in COPD

MMP-9, also referred to as type IV collagenase or gelatinase B, is a proenzyme with a molecular mass of 92 kDa. The MMP-9 gene is on human chromosome 20q11.1–13.1. MMP-9 has been cloned in several mammalian species showing 80% homology at the mRNA level between mouse and human. However, the mouse mRNA has additional base pairs resulting in a proenzyme with a mass of 105 kDa. MMP-9 contains fibronectin type II-like repeats within the catalytic domain, giving it a high affinity to gelatin and to elastin [8]. Comprehensive reviews about biochemical properties, regulation and functions of MMP-9 are available elsewhere [9, 10].

In patients with emphysema, there is an increase in bronchoalveolar lavage concentrations and macrophage expression of various MMPs, including MMP-9 [6]. There is also an increase in activity of MMP-9 in the lung parenchyma of patients with emphysema [11] which is correlated with FEV_1 [12]. Alveolar macrophages from smokers express more MMP-9 than those from non-smokers [13], and there is an even greater increase in cells from patients with COPD [14]. The increased MMP-9 generates greatly enhanced elastolytic activity [15]. MMP-9 and the ratio of MMP-9/TIMP-1 are increased in induced sputum of patients with COPD [16]. Patients with high resolution computed tomography (HRCT) confirmed emphysema were also shown to have more MMP-9 and a higher MMP-9/TIMP-1 ratio in their sputum [17].

However, the precise functions of MMP-9 in the lung remain to be established and it is not yet clear whether it is involved in lung remodelling or is part of the inflammatory lung response and repair process [18].

Involvement of MMP-12 in COPD

MMP-12 is a 54 kDa proenzyme, with a 45 kDa NH2-terminal active form that is processed into a mature 22 kDa form. The human gene, which is designated human macrophage metalloelastase, produces a 1.8 kb transcript encoding a 470-amino acid protein that is 64% identical to the mouse protein. Both the mRNA and protein are detected in alveolar macrophages. As in the mouse, the predicted human 54 kDa protein is processed by loss of both N- and C-terminal residues to the 22 kDa mature form [19].

MMP-12 seems to play a predominant role in the pathogenesis of chronic lung injury and particularly in emphysema. Indeed, MMP-12 is able to degrade different substrates among which is elastin [20]. Elastin represents about 2.5% (wt/wt) of the dry weight of the lung and is distributed widely throughout the lung [21]. This protein is vital for the elastic recoil of the small airways and their ability to resist negative pressure collapse. In emphysema, elastin content of the lung parenchyma is decreased; and ultrastructurally, elastic fibers are disorganised and probably nonfunctional [22]. Moreover, elastin degradation products, such as desmosine, are in greater amount in the urine of subjects with COPD [23] and correlate with the rate of lung function decline [24]. *In vitro* studies on alveolar macrophages collected from COPD patients have shown that they degrade more elastin than macrophages collected from controls [15]. Using immunocytochemistry, we have previously observed MMP-12-positive macrophages in both COPD and control samples [25]. However, the number of MMP-12 expressing-macrophages together with the staining intensity was higher in bronchoalveolar lavage samples from COPD patients than in control subjects. Similar results of higher MMP-12 expression were observed in bronchial biopsies from COPD subjects in comparison with controls. Enhanced MMP-12 activity was also shown in BAL fluids from patients with COPD in comparison with control subjects [25, 26]. These studies demonstrated that COPD patients produce greater quantities of MMP-12 than controls, which may be a critical step in the pathogenesis of COPD and emphysema.

Studies using MMP-12$^{-/-}$ KO mice have demonstrated that macrophage recruitment and emphysema induced by long-term exposure to cigarette smoke were linked to MMP-12 [27]. Indeed, when MMP-12$^{-/-}$ KO mice were subjected to cigarette smoke over a 6 months period, in contrast to wild-type control mice, they did not have increased numbers of macrophages in their lungs and did not develop emphysema [27]. The monthly intratracheal instillation of monocyte chemoattractant protein-1 (MCP-1) in the lungs of smoke-exposed MMP-12$^{-/-}$ KO mice caused an increase in macrophage recruitment. However, despite the presence of the macrophages, these MMP-12$^{-/-}$ KO mice did not develop airspace enlargement in response to smoke exposure. These data suggest that MMP-12 is sufficient for the development of emphysema that results from chronic inhalation of cigarette smoke [27].

The macrophage recruitment observed in response to cigarette smoke could be linked to the elastolytic properties of MMP-12. Indeed, MMP-12 generates elastin-derived peptides and experiments realised in modified Boyden chambers have shown that these elastin fragments had chemotactic properties towards monocytes [28, 29].

In a more recent study, it was reported that inflammatory lesions contained significantly more MMP-12 in macrophages in the lungs of mice at 10, 20 and 30 days of cigarette smoke exposure than in controls exposed for 60 days [30]. These results suggest that elastin degradation took place during the development of pulmonary changes in mice exposed to cigarette smoke and that activation of MMPs specific to elastin may be a determining factor for susceptibility to emphysema.

In an acute model of smoke exposure, neutrophils, desmosine and hydroxyproline, two markers of elastin and collagen breakdown, respectively, were examined in BAL fluids of MMP-12$^{-/-}$ KO mice and wild-type mice 24 h after smoke exposure. None of these markers could be detected in MMP-12$^{-/-}$ KO mice, suggesting that acute smoke-induced connective tissue breakdown, the initial step to emphysema, requires both neutrophils and MMP-12 and that the neutrophil influx in the airways is dependent on the presence of MMP-12 [31]. We recently demonstrated that MMP-12$^{-/-}$ KO mice have a reduced airway inflammatory reaction following exposure of two cigarettes twice a day after 2 days of smoke exposure [32]. Interestingly, we also observed by zymography a dramatic concomitant reduction of MMP-9 activity in BAL fluid, indicating that the increase in MMP-9 activity in response to cigarette smoke depends almost entirely on MMP-12 expression. Moreover, the phosphodiesterase(PDE)-4 inhibitor, cilomilast, was able to block neutrophilia induced by cigarette smoke, whereas the treatment with the corticosteroid dexamethasone was ineffective to do so [32]. In contrast to the observations following cigarette smoke exposure, MMP-12$^{-/-}$ KO mice developed a similar airway neutrophilia as control mice when exposed to lipopolysaccharide (LPS). This suggests clear differences between the two models and that the early inflammatory processes following cigarette smoke or LPS exposure although similar in profile, have different causal mechanisms [32].

We recently reported that MMP-12 does not appear to be involved in the fibrogenic response in the lungs following bleomycin-instillation in the airways. Even though an increased expression of MMP-12 in mice lung following bleomycin, MMP-12 deficiency in mice did not influence the bleomycin-induced rise of either TGF-β-1 or TIMP-1 in lungs, which are both described as important pro-fibrogenic effectors [33]. Moreover, in this model, MMP-12 deficiency had no influence on macrophage accumulation in lungs. In agreement with the protease/antiprotease imbalance hypothesis, accumulation of extracellular matrix may be due to dysfunction of MMP-12. These results however contrast with the recent report that lung fibrosis in mice following repeated Fas ligand activation resulted in a sustained increase in MMP-12 expression and that MMP-12$^{-/-}$ KO mice were resistant to development of fibrosis [34].

The direct effect of MMP-12 in the development of inflammatory processes in mouse airways has also been evaluated using a recombinant form of human MMP-12 (rhMMP-12). A single instillation of rhMMP-12 in mouse airways elicited an intense inflammatory response characterised by the development of two successive phases. Indeed, a marked recruitment of neutrophils was observed following rhMMP-12 injection with a maximum increase at 18 h. This cellular recruitment was associated with a very transient increase in IL-6, TNF-α, MIP-1α, MCP-1 and KC levels and gelatinase expression in BAL fluids and in lung parenchyma. From day 4 to day 15, performing the same analyses, we observed an important and stable recruitment of macrophages in BAL fluids in the absence of the inflammatory markers observed during the early phase of inflammation [35]. As this experimental model of lung inflammation partially mimics some features of COPD, we have investigated the effects of treatment with anti-inflammatory compounds, dexamethasone, the phosphodiesterase inhibitor rolipram and a broad-spectrum matrix metalloproteinase (MMP) inhibitor, marimastat [36]. Marimastat (100 mg/kg), dexamethasone (10 mg/kg) and rolipram (0.1 and 0.3 mg/kg) were able to significantly decrease neutrophil recruitment 4 and 24 h after rhMMP-12 instillation, but only marimastat (30 and 100 mg/kg) was effective at decreasing the macrophage recruitment occurring at day 7. Marimastat (100 mg/kg), dexamethasone (10 mg/kg) and rolipram (0.3 mg/kg) significantly reduced the levels of IL-6, KC/CXCL1, MIP-1α/CCL3 and MMP-9 in BAL fluids. Overall, this suggests that dexamethasone and rolipram were able to inhibit the early inflammatory response but were ineffective to limit the macrophage influx. In contrast, marimastat was able to reduce both early and late responses. Because of its ability to induce an inflammatory response and tissue remodelling, it may be possible to consider MMP-12 as an essential component of the process leading to the development of COPD.

Effect of MMP inhibition in experimental models of COPD

COPD is an unmet medical need and development of selective MMP inhibitors is expected to offer new therapeutic opportunities. However, until now it was difficult to confirm the role of specific MMPs because of the lack of selective inhibitors. The limited studies include guinea pigs that were exposed to cigarette smoke over 1 month, 2 months, and 4 months, and received CP-471,474, a broad spectrum

MMP inhibitor [37]. It was reported that CP-471,474 significantly reduced both the extent and severity of inflammation at 2 months. Moreover, the inhibitor significantly decreased the destructive lesions in the lungs, mainly at 2 and 4 months. In another study [38], two orally bioavailable synthetic inhibitors (RS-113456 and RS-132908) were tested in the murine model of smoke-induced emphysema. Both compounds were potent (<1.2 nM) inhibitors of both human and murine MMP-12. RS-132908 markedly inhibited the smoke-induced increase in emphysema at every 6-week observation and produced 74% inhibition at the end of 6 months of smoke exposure. RS-113456 was also extremely effective since it completely blocked further experimental disease. Interestingly, both compounds also reduced macrophage accumulation within the lung tissue. In a more recent study, Churg et al. [39] examined the effects of a dual MMP-9/MMP-12 inhibitor, AZ11557272, on the development of anatomic and functional changes associated with experimental COPD in guinea pigs exposed daily to cigarette smoke for up to 6 months. At all times, smoke-induced increases in lavage inflammatory cells and desmosine and in serum TNF-α were completely suppressed by AZ11557272. After 6 months, AZ11557272 reverted smoke-induced airspace enlargement by about 70%. This study demonstrates that a MMP-9/MMP-12 inhibitor can substantially impede emphysema development, small airway remodelling and improve the functional consequences of these lesions in a non-murine species. It confirms MMP-9 and MMP-12 as potential targets for therapeutic intervention.

Effect of AS112108 on acute airway inflammation induced by cigarette smoke

We have investigated the effect of another dual MMP-9/MMP-12 inhibitor, AS112108, in an acute model of cigarette smoke-induced lung inflammation. C57/BL6 male mice were orally administered with either the dual MMP-12/MMP-9 inhibitor AS112108 (10, 30, 100, 300 mg/kg) or with vehicle (PEG 400: distilled water [1:1]), 1 h before each exposure to cigarette smoke. Mice were then exposed to the smoke of two cigarettes, twice a day, for 3 consecutive days, as previously described [32].

Exposure to cigarette smoke elicited a marked and significant influx of inflammatory cells in BAL fluids in mice at day 4, as compared to control conditions, i.e., exposure to air. These inflammatory cells were mainly neutrophils ($128 \pm 17 \times 10^3$ compared to $0.9 \pm 0.4 \times 10^3$ in controls) (Fig. 1), whereas the number of macrophages did not change significantly. Interestingly, these two cell types represented the majority of the cells recovered from BAL fluids, since lymphocytes and eosinophils accounted for only a maximum of 0.5% of the total cells. Treatment of the mice with the dual MMP-12/MMP-9 inhibitor AS112108 inhibited the increase in the number of neutrophils in a statistically significant manner at 30, 100 and

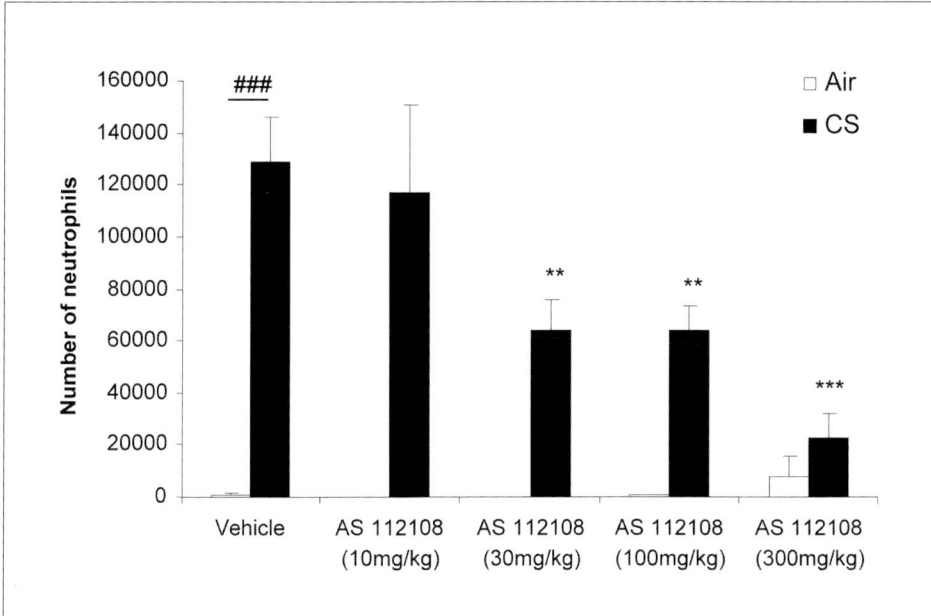

Figure 1
Number of neutrophils in BAL fluids collected at day 4 in mice exposed to cigarette smoke (CS) or to air and administered orally with either vehicle (PEG 400: distilled water, 1:1, v:v) or AS112108 (10, 30, 100, 300 mg/kg) 1 h before cigarette smoke exposure. Results are presented as means ± SEMs and were compared by a two-way analysis of variance (ANOVA). Comparison of treatment interactions was done by Fischer tests. ###: $p < 0.001$ for mice exposed to cigarette smoke compared with mice exposed to air. **: $p < 0.01$, ***: $p < 0.001$ compared with mice exposed to cigarette smoke and administered with vehicle. N = 6–7 animals per group.

300 mg/kg ($p < 0.01$ *versus* vehicle treated, cigarette smoke exposed mice). In contrast, the treatment of mice with AS112108 did not modify the number of neutrophils and macrophages in BAL fluids of mice exposed to air.

By use of zymography of the BAL fluids, the following gelatinolytic bands were identified as metalloproteinase activities: pro-MMP-9 (105 kDa), pro-MMP-2 (72 kDa) and MMP-2 (64 kDa). At day 4, exposure to cigarette smoke induced an increased expression of pro-MMP-9 (105 kDa) in BAL fluids, in comparison with control mice (Fig. 2A). MMP-2 (64 kDa) expression was also increased after cigarette smoke exposure, but to a lesser degree (data not shown). In contrast, in lung parenchyma, no difference in both MMP-9 and MMP-2 levels could be observed between controls and mice exposed to cigarette smoke (data not shown). The treatment of mice exposed to cigarette smoke with AS112108 dose-dependently

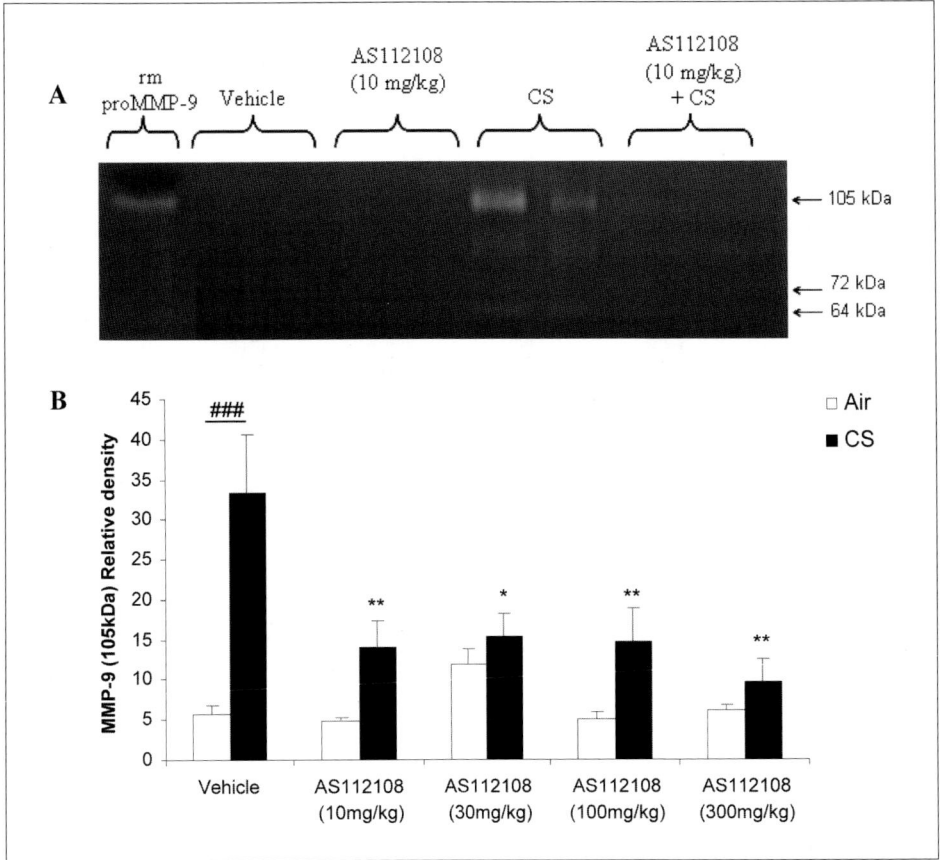

Figure 2
Zymographic analysis of the BAL fluids recovered in the airways of mice exposed to the smoke of two cigarettes twice daily for 3 days. Mice were orally administered with either vehicle (PEG 400: distilled water) or AS112108 (10, 30, 100, 300 mg/kg) 1 h before each cigarette smoke exposure.

(A) Representative gelatin zymogram performed with samples of BAL fluids from mice exposed to air and orally administered with vehicle (lanes 2–3) or with AS112108, 10 mg/kg (lanes 4–5), or from mice exposed to cigarette smoke and administered with vehicle (lanes 6–7) or with AS112108, 10 mg/kg (lanes 8–9). Lane 1 = recombinant mouse pro-MMP-9 (rm-pro-MMP 9).

(B) Quantification by densitometry of 105 kDa gelatinase activity on zymograms of BAL fluids performed at day 4, from mice exposed to cigarette smoke (black bar) or to air (white bar). Results are presented as means ± SEMs and were compared by a two-way analysis of variance (ANOVA). Comparison of treatment interactions was done by Fischer tests. ###: $p < 0.001$ for mice exposed to cigarette smoke compared with mice exposed to air. *: $p < 0.05$, **: $p < 0.01$ compared to mice orally administered with vehicle and exposed to cigarette smoke. N = 6–7 per group.

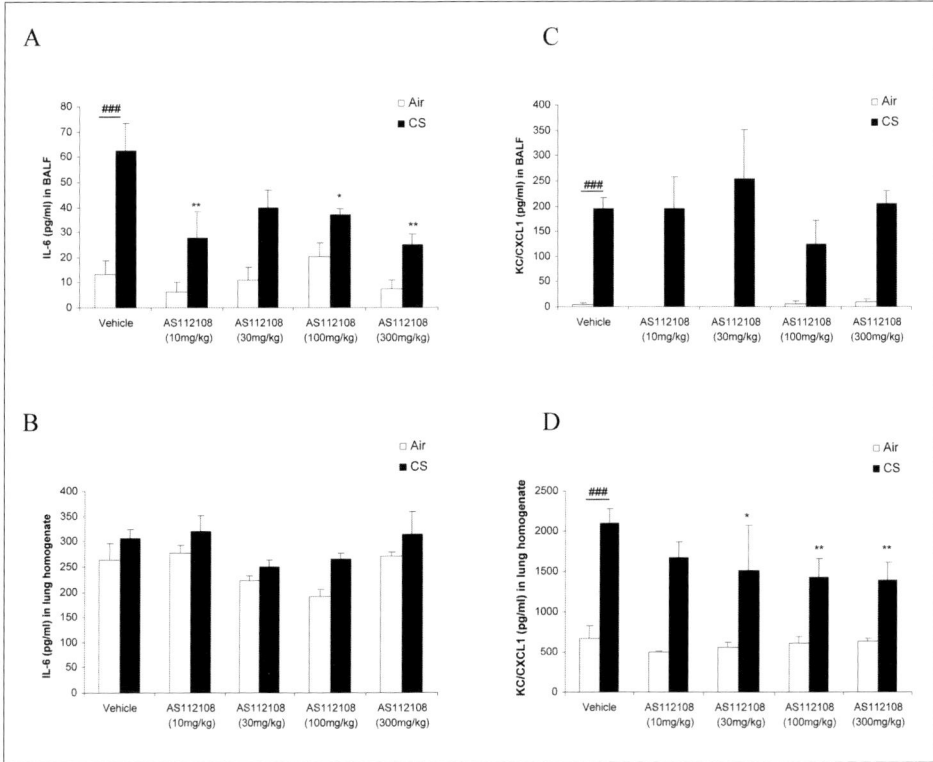

Figure 3
Levels of IL-6 and KC/CXCL-1 in BAL fluids (BALF) (A and C) and in lung homogenate super-natants (B and D) from mice exposed to cigarette smoke (black bar) or to air (white bar) and orally administered with either vehicle (PEG 400: distilled water) or AS112108 (10, 30, 100, 300 mg/kg). BAL fluids and lung tissues were collected 24 h after the last cigarette smoke exposure (on day 4). Results are expressed in pg/ml. Results are presented as means ± SEMs and were compared by a two-way analysis of variance (ANOVA). Comparison of treatment interactions was done by Fischer tests. ###: $p < 0.001$ for mice exposed to cigarette smoke compared with mice exposed to air. **: $p < 0.01$, *: $p < 0.05$ compared to mice orally admin-istered with vehicle and exposed to cigarette smoke. N = 6–7 per group.

inhibited the MMP-9 increased activity in BAL fluids, which appeared statistically significant at 10–300 mg/kg (Fig. 2B).

At day 4, IL-6 concentration significantly increased in BAL fluids of mice exposed to cigarette smoke in comparison with control mice ($p < 0.001$; Fig. 3A). The treatment of cigarette smoke-exposed mice with AS112108 induced an inhibi-tion of IL-6 release in BAL fluids, which was statistically significant at 10, 100 or 300 mg/kg ($p < 0.01$, $p < 0.05$ and $p < 0.01$, respectively). In contrast, despite higher levels of IL-6, there was no significant change with AS112108 (Fig. 3B) in the lung parenchyma.

Exposure of mice to cigarette smoke also induced a significant increase in KC/CXCL1 levels in BAL fluids in comparison to control mice but AS112108 failed to prevent such an increase (Fig. 3C). In contrast, treatment of mice with AS112108 (30, 100 and 300 mg/kg) slightly but significantly ($p < 0.05$ and $p < 0.01$) decreased KC/CXCL1 levels in the lung parenchyma (Fig. 3D).

Concluding remarks

The underlying mechanisms of emphysema are associated with inflammatory and remodelling processes in the airways. Because of their involvement in the inflammatory response and tissue remodelling, it may be possible to consider MMPs as essential components of the process leading to the development of COPD. Hence, MMPs could be pivotal in the two main hypotheses that are proposed to explain the pathological process of COPD, i.e., the 'elastase/antielastase imbalance' theory and the 'inflammation/repair' theory.

Here we presently report that a MMP-9/MMP-12 inhibitor was able to markedly reduce inflammatory processes associated with cigarette smoke exposure, suggesting that this compound may interfere with the early events of the pathophysiological processes associated with COPD. These and other data support the hypothesis that MMP-9 and MMP-12 play predominant roles in the inflammatory process induced by cigarette smoke, and therefore, that both MMPs are potentially important therapeutic targets for the treatment of COPD.

Acknowledgements

The authors thank Pr. Ira Katz for the helpful advice regarding this manuscript.

References

1   Pauwels R (2000) COPD: the scope of the problem in Europe. *Chest* 117: 332S–335S
2   Saetta M (1999) Airway inflammation in chronic obstructive pulmonary disease. *Am J Respir Crit Care Med* 160: S17–20
3   Eidelman D, Saetta MP, Ghezzo H, Wang NS, Hoidal JR, King M, Cosio MG (1990) Cellularity of the alveolar walls in smokers and its relation to alveolar destruction. Functional implications. *Am Rev Respir Dis* 141: 1547–1552
4   Finkelstein R, Fraser RS, Ghezzo H, Cosio MG (1995) Alveolar inflammation and its relation to emphysema in smokers. *Am J Respir Crit Care Med* 152: 1666–1672

5   Tetley TD (2002) Macrophages and the pathogenesis of COPD. *Chest* 121: 156S–159S

6   Finlay GA, O'Driscoll LR, Russell KJ, D'Arcy EM, Masterson JB, FitzGerald MX, O'Connor CM (1997). Matrix metalloproteinase expression and production by alveolar macrophages in emphysema. *Am J Respir Crit Care Med* 156: 240–247

7   Shapiro SD, Senior RM (1999). Matrix metalloproteinases. Matrix degradation and more. *Am J Respir Cell Mol Biol* 20: 1100–1102

8   Shipley JM, Doyle GA, Fliszar CJ, Ye QZ, Johnson LL, Shapiro SD, Welgus HG, Senior RM (1996) The structural basis for the elastolytic activity of the 92-kDa and 72-kDa gelatinases. Role of the fibronectin type II-like repeats. *J Biol Chem* 271: 4335–4341

9   Vu T, Werb Z (1998). Gelatinase B: structure, regulation, and function. In: WC Parks, RP Mecham (eds): *Matrix Metalloproteinases.* Academic Press, San Diego, 115–148

10   Van den Steen PE, Dubois B, Nelissen I, Rudd PM, Dwek RA, Opdenakker G (2002) Biochemistry and molecular biology of gelatinase B or matrix metalloproteinase-9 (MMP-9). *Crit Rev Biochem Mol Biol* 37: 375–536

11   Ohnishi K, Takagi M, Kurokawa Y, Satomi S, Konttinen YT (1998) Matrix metalloproteinase-mediated extracellular matrix protein degradation in human pulmonary emphysema. *Lab Invest* 78: 1077–1087

12   Kang MJ, Oh YM, Lee JC, Kim DG, Park MJ, Lee MG, Hyun IG, Han SK, Shim YS, Jung KS (2003) Lung matrix metalloproteinase-9 correlates with cigarette smoking and obstruction of airflow. *J Kor Med Sci* 18: 821–827

13   Lim S, Roche N, Oliver BG, Mattos W, Barnes PJ, Chung KF (2000) Balance of matrix metalloprotease-9 and tissue inhibitor of metalloprotease-1 from alveolar macrophages in cigarette smokers. Regulation by interleukin-10. *Am J Respir Crit Care Med* 162: 1355–1360

14   Russell RE, Culpitt SV, DeMatos C, Donnelly L, Smith M, Wiggins J, Barnes PJ (2002) Release and activity of matrix metalloproteinase-9 and tissue inhibitor of metalloproteinase-1 by alveolar macrophages from patients with chronic obstructive pulmonary disease. *Am J Respir Cell Mol Biol* 26: 602–609

15   Russell RE, Thorley A, Culpitt SV, Dodd S, Donnelly LE, Demattos C, Fitzgerald M, Barnes PJ (2002) Alveolar macrophage-mediated elastolysis: roles of matrix metalloproteinases, cysteine, and serine proteases. *Am J Physiol Long Cell Mol Physiol* 283: L867–L873

16   Cataldo D, Munaut C, Noel A, Frankenne F, Bartsch P, Foidart JM, Louis R (2000) MMP-2- and MMP-9-linked gelatinolytic activity in the sputum from patients with asthma and chronic obstructive pulmonary disease. *Int Arch Allergy Immunol* 123: 259–267

17   Boschetto P, Quintavalle S, Zeni E, Leprotti S, Potena A, Ballerin L, Papi A, Palladini G, Luisetti M, Annovazzi L et al (2006) Association between markers of emphysema and more severe chronic obstructive pulmonary disease. *Thorax* 61: 1037–1042

18   Atkinson JJ, Senior RM (2003) Matrix metalloproteinase-9 in lung remodeling. *Am J Respir Cell Mol Biol* 28: 12–24

19 Shapiro SD, Kobayashi DK, Ley TJ (1993). Cloning and characterization of a unique elastolytic metalloproteinase produced by human alveolar macrophages. *J Biol Chem* 268: 23824–23829

20 Gronski TJ, Martin RL, Kobayashi DK, Walsh BC, Holman M, Huber M, Van Wart HE, Shapiro SD (1997) Hydrolysis of a broad spectrum of extracellular matrix proteins by human macrophage elastase. *J Biol Chem* 272: 12189–12194

21 Starcher BC (1986) Elastin and the lung. *Thorax* 41: 577–585

22 Shapiro SD (2000) Evolving concepts in the pathogenesis of chronic obstructive pulmonary disease. *Clinical Chest Medical* 21: 621–632

23 Stone PJ, Gottlieb DJ, O'Connor GT, Ciccolella DE, Breuer R, Bryan-Rhadfi J, Shaw HA, Franzblau C, Snider GL (1995) Elastin and collagen degradation products in urine of smokers with and without chronic obstructive pulmonary disease. *Am J Respir Crit Care Med* 151: 952–959

24 Gottlieb DJ, Stone PJ, Sparrow D, Gale ME, Weiss ST, Snider GL, O'Connor GT (1996) Urinary desmosine excretion in smokers with and without rapid decline of lung function: the Normative Aging Study. *Am J Respir Crit Care Med* 154: 1290–1295

25 Molet S, Belleguic C, Léna H, Germain N, Bertrand CP, Shapiro SD, Planquois JM, Delaval P, Lagente V (2004) Increase in macrophage elastase (MMP-12) in lungs from patients with chronic obstructive pulmonary disease. *Inflamm Res* 54: 31–36

26 Demedts IK, Morel-Montero A, Lebecque S, Pacheco Y, Cataldo D, Joos GF (2006) Elevated MMP-12 protein levels in induced sputum from patients with COPD. *Thorax* 61: 196–201

27 Hautamaki RD, Kobayashi DK, Senior RM, Shapiro SD (1997). Requirement for macrophage elastase for cigarette smoke-induced emphysema in mice. *Science* 277: 2002–2004

28 Senior RM, Griffin GL, Mecham RP (1980) Chemotactic activity of elastin-derived peptides. *J Clin Invest* 66: 859–862

29 Houghton AM, Quintero PA, Perkins DL, Kobayashi DK, Kelley DG, Marconcini LA, Mecham RP, Senior RM, Shapiro SD (2006) Elastin fragments drive disease progression in a murine model of emphysema. *J Clin Invest* 116: 753–759

30 Valença SS, Da Hora K, Castro P, Gonçalves de Moraes V, Carvalho L, Moraes Sobrino Porto LC (2004) Emphysema and metalloelastase expression in mouse lung induced by cigarette smoke. *Tox Pathol* 32: 351–356

31 Churg A, Zay K, Shay S, Xie C, Shapiro SD, Hendricks R, Wright JL (2002) Acute cigarette smoke-induced connective tissue breakdown requires both neutrophils and macrophage metalloelastase in mice. *Am J Respir Cell Mol Biol* 27: 368–374

32 Leclerc O, Lagente V, Planquois JM, Berthelier C, Artola M, Eichholtz T, Bertrand CP, Schmidlin F (2006) Involvement of MMP-12 and phosphodiesterase type 4 in cigarette smoke-induced inflammation in mice. *Eur Respir J* 27: 1102–1109

33 Manoury B, Nenan S, Guenon I, Boichot E, Planquois JM, Bertrand CP, Lagente V (2006) Macrophage metalloelastase (MMP-12) deficiency does not alter bleomycin-induced pulmonary fibrosis in mice. *J Inflamm (Lond)* 22: 2

34 Matute-Bello G, Wurfel MM, Lee JS, Park DR, Frevert CW, Madtes DK, Shapiro SD, Martin TR (2007) Essential role of MMP-12 in Fas-induced lung fibrosis. *Am J Respir Cell Mol Biol* 37: 210–221

35 Nenan S, Planquois JM, Berna P, De Mendez I, Hitier S, Shapiro SD, Boichot E, Lagente V, Bertrand CP (2005) Analysis of the inflammatory response induced by rhMMP-12 catalytic domain instilled in mouse airways. *Int Immunopharmacol* 5: 511–524

36 Nenan S, Lagente V, Planquois JM, Hitier S, Berna P, Bertrand CP, Boichot E (2007) Metalloelastase (MMP-12) induced inflammatory response in mice airways: effects of dexamethasone, rolipram and marimastat. *Eur J Pharmacol* 559: 75–81

37 Selman M, Cisneros-Lira J, Gaxiola M, Ramirez R, Kudlacz EM, Mitchell PG, Pardo A (2003) Matrix metalloproteinases inhibition attenuates tobacco smoke-induced emphysema in Guinea pigs. *Chest* 123: 1633–1641

38 Martin RL, Shapiro SD, Tong SE, Van Wart HE (2001) Macrophage metalloelastase inhibitors. *Prog Respir Res* 31: 177–180

39 Churg A, Wang R, Wang X, Onnervik PO, Thim K, Wright JL (2007) Effect of an MMP-9/MMP-12 inhibitor on smoke-induced emphysema and airway remodeling in guinea pigs. *Thorax* 62: 706–713

Role of matrix metalloproteinases (MMPs) in cystic fibrosis

Stéphane Jouneau[1,2], Guillaume Léveiller[1], Sylvie Caulet-Maugendre[3], Graziella Brinchault[1], Chantal Belleguic[1], Benoît Desrues[1] and Vincent Lagente[2]

[1]Respiratory Medicine Department, Pontchaillou Hospital, 35000 Rennes, France; [2]INSERM U620, Faculté de Pharmacie, Université de Rennes 1, 2 avenue du Professeur Léon Bernard, 35000 Rennes, France; [3]Pathology Department, Pontchaillou Hospital, 35000 Rennes, France

Abstract

Cystic fibrosis (CF) is the most frequent hereditary lethal disease in Caucasians. It is the consequence of a mutation in a chloride channel called CFTR. This defective channel leads to viscous secretions in all exocrine glands and therefore destruction and fibrosis of these organs. The main point of CF is the lung disease with inflammation and infection leading towards remodelling. There are only symptomatic treatments. Matrix metalloproteinases (MMPs) play several roles in CF development. Indeed, MMPs are involved in the regulation of CFTR channel. The alveolar levels of MMPs in CF patients are increased compared to controls with active form and lead to an imbalance between proteases and anti-proteases. MMPs are enhanced in sputum and plasma in severe CF patients. MMPs also have a role in regeneration of human CF airway surface epithelium and differentiation. MMPs could also interfere with the aerosolised medication. Together, these data exhibit the major role of MMPs in CF.

Introduction

Matrix metalloproteinases are a family of more than 20 enzymes which can degrade all the components of the extracellular matrix. There are four main families: collagenases, gelatinases (MMP-2 and MMP-9), membrane-type (MT)-MMP-(1 to 4) and stromelysins. MMPs have been reported to be involved in tissue remodelling and in inflammatory processes, two major patterns present in cystic fibrosis. Therefore, the aim of the present chapter is to analyse the involvement of MMPs in the development and the evolution of cystic fibrosis.

Cystic fibrosis (CF)

Physiopathology

Cystic fibrosis (CF) is the most frequent lethal hereditary disease in caucasians. CF is due to a mutation in the CFTR (cystic fibrosis transmembrane conductance regulator) gene localised in 7q31.2 [1]. This gene is encoding for a transmembrane protein also called CFTR. The CFTR protein belongs to ABC (ATP binding cassette) protein family and its main role is as an outward chloride channel on the apical pole of many epithelial cells in the organism especially ciliated cells in proximal and distal airways. Because of the deficient CFTR protein, the epithelial sodium channel (ENaC) is not inhibited leading to a hyperabsorption in sodium and therefore in water, then the secretions become too viscous. In the lung, the mucociliary escalator is not working properly, secretions are blocked and become infected. This blockage by viscous secretions occurs also in the other organs and drives toward remodelling (fibrosis and destruction) in all exocrine glands. But CFTR has other roles such as transferring other molecules (bicarbonate, urea, water), regulating other channels (chloride channel, potassium channel etc.), regulating pH of intracytoplasmic organelles, regulating exocytosis–endocytosis and apoptosis etc. [1–6].

Clinical features

The clinical features are dominated by the respiratory disease. CF is characterised by bronchiectasis, mainly in the apex of both lungs, which occurs most of the time during childhood. Chronic cough and purulent sputum are the main signs of the disease. The evolution is characterised by recurrent infectious exacerbations and sometimes life-threatening complications such as pneumothorax or haemoptysis. Bacterial colonisation is usual and crucial, primarily with *Staphyloccocus aureus* and *Haemophilus influenzae*, and later *Pseudomonas aeruginosa*, signing a clear degradation in their health and requiring a combination of antibiotics to treat the infectious exacerbations. We observed a progressive deterioration driving toward chronic respiratory failure which can be assessed by a functional respiratory test: forced expiratory volume in one second (FEV1) and forced vital capacity (FVC) but also with the X-ray findings. The severity of the CF lung disease can be assessed by a radiological score based on the chest X-ray, the Brasfield score [7, 8]. This score assesses the air trapping, linear marking, nodular cystic lesions, large lesions and general severity (25 = excellent, 5 = severe).

The second type of manifestations are digestive. There is an exocrine pancreatic failure with chronic steatorrhea and possibly an endocrine pancreatic incapacity secondary to the fibrosis of the pancreas leading to diabetes. Liver is also affected, from initial cholestasis to final cirrhosis. The combination of respiratory and digestive disorders drives toward an alteration of the general health which could be assessed by the Shwachman score (citation of the general activity, physical activity, nutrition and X-ray findings, 100 = excellent, 20 = very severe) [9].

We also observed different manifestations of the disease: nasal polyposis and chronic rhinosinusitis, osteoporosis and genital abnormalities in both sex (agenesis of vas deferens and abnormal cervical secretions).

Diagnosis

The neonatal screening uses the assay of immunoreactive trypsin on dry blood spots. In clinically evocating situation, the sweat chloride concentration by pilocarpine iontophoresis has been evaluated twice. In CF, this chloride concentration is > 60 mmol/L. In any cases, the CF diagnosis needs to be confirmed with CFTR gene mutations research [10].

Treatment

There is no curative treatment at this moment but several treatments are proposed to reduce symptoms. The main point of the respiratory treatment is the daily chest physiotherapy, associated with the Rh-DNAse aerosol in some responder patients [11]. During the recurrent infectious exacerbations, intravenous antibiotics are essential. In chronic infection, antibiotics can be delivered in aerosol. If there is a reversible part of the functional respiratory tests, long acting beta-agonists can be useful. Oxygenotherapy and possibly lung transplantation could be indicated.

From a digestive point of view with exocrin pancreatic incapacity, pancreatic enzyme extracts are prescribed [12]. Vitamins and oligoelement supplementation are indicated accordingly to the serum level. The diet should be enriched in protein and energy. If a biological cholangitis appeared, ursodeoxycholic acid is prescribed. Eventually, the liver transplantation could be indicated. The other symptoms are specifically followed such as diabetes, sterility, etc.

Involvement of inflammation in cystic fibrosis

There are two pathways of inflammatory process in CF: exogenous and endogenous [13]. The exogenous pathway is the consequence of bacterial colonisation in the airways leading to the release of proinflammatory cytokines and proteases. The endogenous pathway appears the most important and seems to be directly linked to CFTR dysfunction. Indeed, we observed the development of early inflammation without sign of infection in bronchoalveolar lavage (BAL) from newborns with CF

[14]. These endogenous processes are associated with a decrease in anti-inflammatory cytokines (namely interleukin (IL)-10) and increase in pro-inflammatory cytokines (such as IL-6, IL-8). We also observed an intense neutrophil sequestration in bronchi from CF patients [15].

MMPs in cystic fibrosis

Proteases may play a central role in the lungs of CF patients. Indeed, human neutrophil elastase (HNE) is a major serine protease which is described to be increased in BAL from CF patients leading to an imbalance between proteases and antiproteases [16]. Because of their ability to degrade the totality of the components of the extracellular matrix, MMPs also play a pivotal role in the remodelling of the airways and inflammation.

Besides the capacity to be involved in the development of the tubulo-acinar glands from the airways, which participate in lung host defence, MMP-2 has been shown to be associated with the regulation of ion channel function [17]. In a submucosal cell line from human airway (Calu-3), Duszyk et al. reported that a specific MMP inhibitor (phenantroline) increased the whole-cell current [18]. Current stimulated by phenantroline displayed linear current-voltage relationships and had inhibitor pharmacology and ion selectivity consistent with CFTR Cl⁻ channel. The Calu-3 cells were expressing MMP-2 assessed by zymography and western blotting. Moreover, anti-MMP-2 antibodies increased whole-cell current, whereas human recombinant MMP-2 reduced it, suggesting that MMP-2 is involved in the regulation of CFTR Cl⁻ channel in human airways. The same experiments have been done on A549 cells, which are derived from airway surface epithelium but do not express CFTR Cl⁻ channel. While these cells also show significant expression of MMP-2, inhibition of this enzyme with phenantroline exerted no significant effect on the whole-cell current.

MMP-8, MMP-9 and the ratio MMP-9/TIMP-1 (tissue inhibitor of MMP) were significantly higher in the BAL fluids from children with CF compared to controls, presenting normal lung function and free of exacerbations [19]. The levels of MMP-8 and MMP-9 in BAL fluid were significantly correlated with the percentage of neutrophils in BAL fluids. Active MMP-9 was also found in most CF children but not in controls. A second bronchoscopy was performed 18 months after the first one and the BAL fluid levels of MMP-8 and MMP-9 were increased in CF patients, whereas both MMP-8 and MMP-9 were significantly decreased in DNAse treated patients [19].

Several works have reported similar results in sputum from children with CF. Higher levels of MMP-2, MMP-8, MMP-9, TIMP-1 in sputum and increased molar ratio MMP-9/TIMP-1 have been shown, compared to control groups [20–23]. The active form of MMP-8 and MMP-9 were present mainly in CF sputum compared to control groups [20–23]. The authors found significant correlations between MMPs measurements in sputum such as MMP-8 activity, MMP-9 levels and uninhibited gelatinolytic activity and CF severity assessed by Shwachman score, FEV1 and FVC. Interestingly, bacterial proteinases from *Pseudomonas aeruginosa* slightly contributed to the gelatinolytic activity in CF sputum [20]. In addition, the levels of MMP-9 and TIMP-1 were correlated to lung inflammation, mainly neutrophil count, IL-8 and HNE but not to airway infection [23]. These two observations support the importance of the endogenous inflammation pathway in airway remodelling in CF. Using *in vitro* studies, it has been reported that HNE can activate pro-MMP-9 to the active form and degrades TIMP-1 [21]. This activation was not observed with MMP-8. The strong correlation between HNE sputum level and MMP-9 activity in CF patients probably contributes to the mechanism of the imbalance between proteases and antiproteases in lungs from CF patients [21].

We were interested in extracellular matrix metalloproteinase inducer (EMMPRIN or CD147) which has been described to stimulate the production of MMPs and has never been studied in CF. The aim of our study was to assess the presence of EMMPRIN in CF patients and determine the relationship with the severity of the disease [24]. In the sputum of adult patients, we have noted the presence of EMMPRIN and, as previous works described above, confirmed the proteases/antiproteases imbalance with a mean ratio MMP-9/TIMP-1 > 100. In the plasma of stable CF patients including children and adults without modification of the symptoms and/or above 3 months after a pulmonary infectious exacerbation, the levels of EMMPRIN and MT1-MMP (membrane type 1-MMP) were positively correlated with the severity criteria (FEV1, FVC, Shwachman and Brasfield scores) whereas the level of pro-MMP-2 was inversely correlated. In lung parenchyma, using immunochemistry, we demonstrated that the main sites of EMMPRIN expression were localised in alveolar and tissue macrophages as well as bronchiolar and alveolar type II epithelial cells. We also found a co-localisation of EMMPRIN, MMP-2 and MT1-MMP on tissue macrophages and myofibroblasts in ulcerative inflammatory areas with wound healing. This co-localisation suggests a role of EMMPRIN in severe CF evolution. It is therefore possible that the increase of EMMPRIN would induce an increased production/synthesis of MT1-MMP. Subsequently, MT1-MMP would induce the activation of pro-MMP-2 in active MMP-2, leading to the decrease of plasma level of pro-MMP-2 with the increase of the severity. This active MMP-2 would be only present in the lung parenchyma because we did not find any active MMP-2 in plasma. This active MMP-2 may play a role in the lung destruction and fibrosis (Fig. 1). Our immunohistochemistry data strongly support this hypothesis with the co-localisation of EMMPRIN, MMP-2 and MT1-MMP in the tissue macrophages and myofibroblasts from inflammatory wound healing areas. EMMPRIN presents also an intense staining on alveolar macrophages and epithelial cells in the lungs from CF patients (Fig. 2).

Cystic fibrosis (CF) at an advanced stage of the disease is characterised by airway epithelial injury and remodelling. Puchelle and co-workers have established

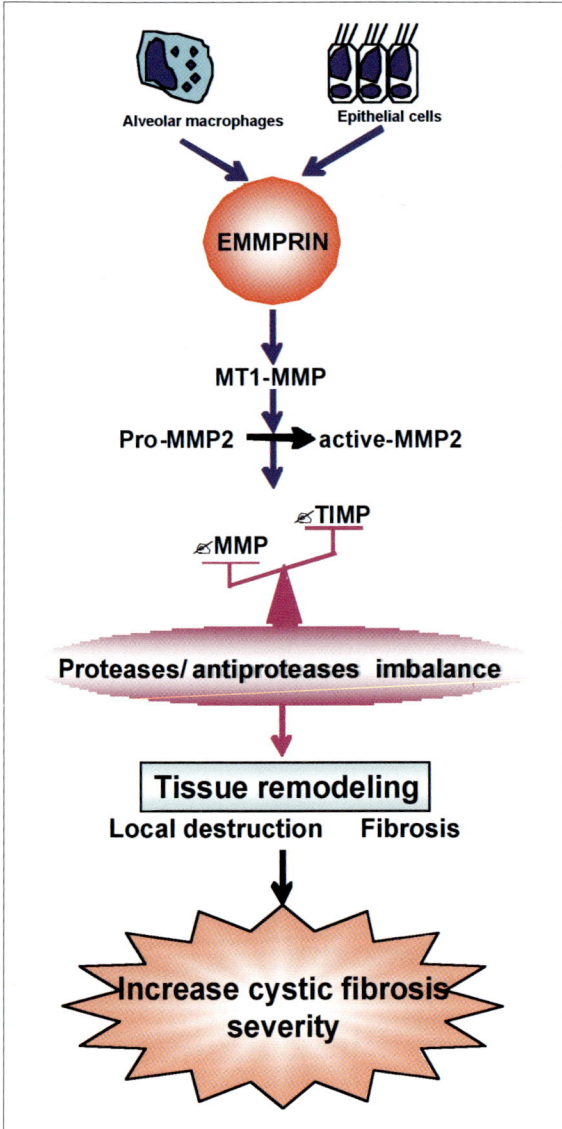

Figure 1
Role of EMMPRIN in the development of cystic fibrosis symptoms.

the expression and secretion profiles of IL-8, MMP-7, MMP-9, and TIMP-1 during non-CF airway epithelial regeneration in a humanised nude mouse xenograft model [25]. To enhance the understanding of CF remodelling, they compared the

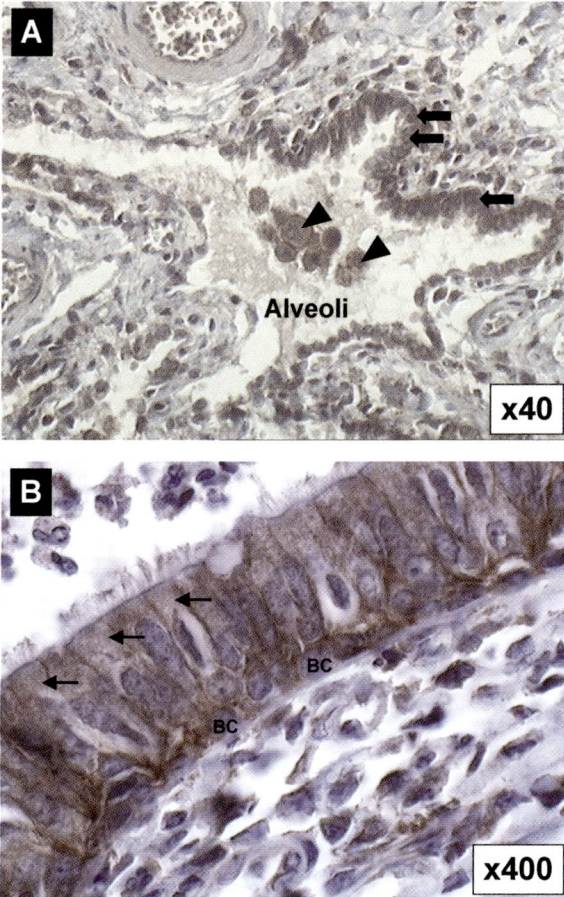

Figure 2
Immunohistochemistry on CF lungs. Strepatvidine-biotin-peroxydase technic against EMMPRIN: A. Strong expression on alveolar macrophages (arrow head) and activated type II pneumocytes (bold arrow). B. Strong expression on bronchiolar ciliated cells (thin arrow) and basal epithelial cells (BC).

regeneration process of non-infected human CF and non-CF nasal epithelia. In both CF and non-CF situations, epithelial regeneration was characterised by successive steps of cell adhesion and migration, proliferation, pseudostratification, and terminal differentiation [26]. However, histological examination of the grafts showed a delay in differentiation of the CF airway epithelium, cell proliferation was higher in the regenerating CF epithelium, and the differentiated CF epithelium exhibited a pronounced height increase and basal cell hyperplasia in comparison with non-CF epithelium. In addition, while the number of goblet cells expressing MUC5AC was similar in CF and non-CF regenerated epithelia, the number of MUC5B-immunopositive goblet cells was lower in CF grafts. The expression of human IL-8, MMP-7, MMP-9, and TIMP-1 was enhanced in CF epithelium, mainly early in the regenerative process. Together, these data strongly suggest that the regeneration of human CF airway surface epithelium is characterised by remodelling, delayed differentiation, and altered pro-inflammatory and MMP responses.

Another aspect of MMPs involvement in CF is their interaction with the local medications. Indeed, as described above, there is an excess in proteases in CF lungs. Therefore, aerosolised treatments in CF have to resist proteolysis, so it is likely to encounter at inflammatory sites *in vivo*. It was one of the main points of the EPI-hNE4 (depelstat), a potent inhibitor of HNE derived from human inter-alpha-trypsin inhibitor and designed to control the excess proteolytic activity in the sputum of CF patients [27]. EPI-hNE4 resisted hydrolysis by neutrophil MMPs and serine proteases that are released from activated neutrophils in inflammatory lung secretions, including MMP-8 and MMP-9, and the elastase-related protease 3 and cathepsin G. It also resisted degradation by epithelial lung cell MMP-7. The authors concluded that EPI-hNE4 seems to be an elastase inhibitor suitable for use in aerosols to treat patients with cystic fibrosis.

Conclusion

MMPs play a central role in CF in different ways. MMPs seem to participate in the regulation of CFTR ion channel function. MMPs are also involved in inflammatory and remodelling processes in CF lungs. Many studies demonstrate a huge imbalance between MMPs and TIMPs, which can be reduce by some of the CF medications such as antibiotics and DNAse. Elevated EMMPRIN in serum is correlated with the severity of CF and co-localise with MMPs in the most inflammatory area of CF lungs. The CF epithelia, expressing more MMPs and IL-8 than non-CF epithelia, presented an increased proliferation but a delay in differentiation. The high level of MMPs in CF lungs may alter the aerosolised medications in these patients. A better understanding of these different processes could improve the management of CF patients and help to find new medications, especially those which can modify the imbalance between MMPs and TIMPs leading to a normalisation of the remodelling process.

References

1   Quinton PM (1999) Physiological basis of cystic fibrosis: a historical perspective. *Physiol Rev* 79 (1 Suppl): S3–S22

2 Barasch J, Kiss B, Prince A, Saiman L, Gruenert D, al-Awqati Q (1991) Defective acidi-fication of intracellular organelles in cystic fibrosis. *Nature* 352(6330): 70–73

3 Hasegawa H, Skach W, Baker O, Calayag MC, Lingappa V, Verkman AS (1992) A multifunctional aqueous channel formed by CFTR. *Science* 258(5087): 1477–1479

4 Loussouarn G, Demolombe S, Mohammad-Panah R, Escande D, Baro I (1996) Expres-sion of CFTR controls cAMP-dependent activation of epithelial K$^+$ currents. *Am J Physiol* 271(5 Pt 1): C1565–C1573

5 Pilewski JM, Frizzell RA (1999) Role of CFTR in airway disease. *Physiol Rev* 79 (1 Suppl): S215–S255

6 Stutts MJ, Canessa CM, Olsen JC, Hamrick M, Cohn JA, Rossier BC, Boucher RC (1995) CFTR as a cAMP-dependent regulator of sodium channels. *Science* 269(5225): 847–850

7 Brasfield D, Hicks G, Soong S, Tiller RE (1979) The chest roentgenogram in cystic fibrosis: a new scoring system. *Pediatrics* 63(1): 24–29

8 Brasfield D, Hicks G, Soong S, Peters J, Tiller R (1980) Evaluation of scoring system of the chest radiograph in cystic fibrosis: a collaborative study. *AJR Am J Roentgenol* 134(6): 1195–1198

9 Shwachman H, Kulczycki LL (1958) Long-term study of one hundred five patients with cystic fibrosis; studies made over a five- to fourteen-year period. *AMA J Dis Child* 96(1): 6–15

10 Comeau AM, Accurso FJ, White TB, Campbell PW III, Hoffman G, Parad RB, Wilfond BS, Rosenfeld M, Sontag MK, Massie J et al (2007) Guidelines for implementation of cystic fibrosis newborn screening programs: Cystic Fibrosis Foundation workshop report. *Pediatrics* 119(2): e495–e518

11 Wagener JS, Headley AA (2003) Cystic fibrosis: current trends in respiratory care. *Respir Care* 48(3): 234–245

12 Littlewood JM, Wolfe SP, Conway SP (2006) Diagnosis and treatment of intestinal mal-absorption in cystic fibrosis. *Pediatr Pulmonol* 41(1): 35–49

13 Derelle J (2003) Airway inflammation in cystic fibrosis. *Rev Prat* 53(2): 141–144

14 Khan TZ, Wagener JS, Bost T, Martinez J, Accurso FJ, Riches DW (1995) Early pulmo-nary inflammation in infants with cystic fibrosis. *Am J Respir Crit Care Med* 151(4): 1075–1082

15 De Rose V (2002) Mechanisms and markers of airway inflammation in cystic fibrosis. *Eur Respir J* 19(2): 333–340

16 Birrer P, McElvaney NG, Rudeberg A, Sommer CW, Liechti-Gallati S, Kraemer R, Hub-bard R, Crystal RG (1994) Protease-antiprotease imbalance in the lungs of children with cystic fibrosis. *Am J Respir Crit Care Med* 150(1): 207–213

17 Infeld MD (1997) Cell-matrix interactions in gland development in the lung. *Exp Lung Res* 23(2): 161–169

18 Duszyk M, Shu Y, Sawicki G, Radomski A, Man SF, Radomski MW (1999) Inhibition of matrix metalloproteinase MMP-2 activates chloride current in human airway epithelial cells. *Can J Physiol Pharmacol* 77(7): 529–535

19   Ratjen F, Hartog CM, Paul K, Wermelt J, Braun J (2002) Matrix metalloproteases in BAL fluid of patients with cystic fibrosis and their modulation by treatment with dornase alpha. *Thorax* 57(11): 930–934

20   Delacourt C, Le BM, d'Ortho MP, Doit C, Scheinmann P, Navarro J, Harf A, Hartmann DJ, Lafuma C (1995) Imbalance between 95 kDa type IV collagenase and tissue inhibitor of metalloproteinases in sputum of patients with cystic fibrosis. *Am J Respir Crit Care Med* 152(2): 765–774

21   Gaggar A, Li Y, Weathington N, Winkler M, Kong M, Jackson P, Blalock JE, Clancy J (2007) Matrix metalloprotease-9 dysregulation in lower airway secretions of cystic fibrosis patients. *Am J Physiol Lung Cell Mol Physiol* 293: L96–L104

22   Power C, O'Connor CM, MacFarlane D, O'Mahoney S, Gaffney K, Hayes J, FitzGerald MX (1994) Neutrophil collagenase in sputum from patients with cystic fibrosis. *Am J Respir Crit Care Med* 150(3): 818–822

23   Sagel SD, Kapsner RK, Osberg I (2005) Induced sputum matrix metalloproteinase-9 correlates with lung function and airway inflammation in children with cystic fibrosis. *Pediatr Pulmonol* 39(3): 224–232

24   Jouneau S, Leveiller G, Desrues B, Lagente V, Martin-Chouly C (2005) Increased EMMPRIN and MT1-MMP levels in the plasma of the stable adult patients with cystic fibrosis. *Eur Respir J* 26 (Suppl 49): 404s

25   Puchelle E, Le SP, Hajj R, Coraux C (2006) Regeneration of injured airway epithelium. *Ann Pharm Fr* 64(2): 107–113

26   Hajj R, Lesimple P, Nawrocki-Raby B, Birembaut P, Puchelle E, Coraux C (2007) Human airway surface epithelial regeneration is delayed and abnormal in cystic fibrosis. *J Pathol* 211(3): 340–350

27   Attucci S, Gauthier A, Korkmaz B, Delepine P, Martino MF, Saudubray F, Diot P, Gauthier F (2006) EPI-hNE4, a proteolysis-resistant inhibitor of human neutrophil elastase and potential anti-inflammatory drug for treating cystic fibrosis. *J Pharmacol Exp Ther* 318(2): 803–809

MMPs, inflammation and pulmonary arterial hypertension

Marie-Pia d'Ortho

INSERM, Unité U841, IRBM, Département Foie-Coeur-Poumon, équipe 6;
and Université Paris 12, Faculté de Médecine, IFR10; and AP-HP, Groupe Henri Mondor
– Albert Chennevier, Service de Physiologie – Explorations Fonctionnelles, 94000 Créteil,
France

Abstract

Pulmonary arterial hypertension (PAH) is characterised by remodelling of small pulmonary arteries leading to a progressive increase in pulmonary vascular resistance and right ventricular failure [1]. PAH can be idiopathic, familial, or associated with a number of conditions or diseases, such as connective tissue disease. Its prognosis is poor, less than 3 yr from diagnosis. The aetiology of severe unexplained pulmonary hypertension remained largely unknown until a few years ago. The gene underlying familial PAH was identified in 2000, the *BMPR-2* gene. However its mutations are not always present, and it probably does not explained the full scope of the disease. PAH is associated with structural alterations in pulmonary arteries including intimal fibrosis, medial hypertrophy and adventitial changes, pointing towards extracellular matrix remodelling which have raised the question of involvement of the matrix degrading enzymes. Among them, serine proteases, such as plasmina and endogenous vascular elastase (EVE), and matrix metalloproteases have been studied. In experimental models, the three of them are increased. Accordingly, their inhibition has preventing and in some cases therapeutic effects. However it should be stressed that opposite consequence of protease inhibition on PAH can be observed depending on the experimental model, either chronic hypoxia-induced PAH (deleterious) or toxic moncrotalin-induced PAH (positive). In humans, only sparse reports exist, that found increase in the MMP inhibitor, TIMP-1, and MMP-2 expression and decreased collagenase (MMP-1). Inflammation is part of the PAH, and accordingly, protease production is a well known part of the inflammatory response. Answering the question whether protease inhibition might represent a therapeutic option in human PAH is however certainly too early.

Introduction

Pulmonary arterial hypertension (PAH) is characterised by remodelling of the small pulmonary arteries leading to a progressive increase in pulmonary vascular resistance and right ventricular failure [1]. PAH can be idiopathic, familial, or associated with a number of conditions or diseases, such as connective tissue disease, congenital heart disease, portal hypertension, HIV infection, and exposure to toxins

Matrix Metalloproteinases in Tissue Remodelling and Inflammation,
edited by Vincent Lagente and Elisabeth Boichot
© 2008 Birkhäuser Verlag Basel/Switzerland

and drugs, including appetite suppressants [1, 2]. Severe unexplained idiopathic pulmonary arterial hypertension (IPAH), previously known as primary pulmonary hypertension, is a rare condition with an estimated incidence of 2–3 per million per year [3]. Its prognosis is poor, less than 3 yr from diagnosis before the advent of modern therapies. More recently, targeted therapy with endothelin receptor antagonists, phosphodiesterase inhibition, and prostanoids have been reported to improve symptoms of breathlessness and, in the case of epoprostenol, survival [4].

In most patients, the condition is believed to evolve over several years, with an initial asymptomatic increase in pulmonary arteriolar reactivity and remodelling. Signs and symptoms appear when the mean pulmonary artery pressure is in the range of 30–40 mmHg at rest (normal is < 20 mmHg). Gradual clinical deterioration occurs when the mean pulmonary artery pressure plateaus 60–70 mmHg and cardiac output progressively declines.

The aetiology of severe unexplained pulmonary hypertension remained largely unknown until a few years ago. Reports of a causal association between appetite-suppressant drugs [5] and the occurrence of severe pulmonary hypertension provided some insight into its pathogenesis. However, the identification of the gene underlying familial PAH (FPAH) in 2000 provided a firm basis for mechanistic studies [6–8]. After localisation of the disease gene to the long arm of chromosome 2 (2q33) [9], two independent groups identified heterozygous germline mutations in the bone morphogenetic protein (BMP) type II receptor (BMPR-II), a receptor for the transforming growth factor (TGF)-β superfamily [7, 8]. Mutations in the *BMPR2* gene have been found in approximately 70% of families [10]. In addition, up to 25% of patients with apparently sporadic IPAH have been found to harbour similar mutations [11].

PAH that occurs in association with collagen vascular disease or congenital left-to-right shunting and may be triggered by appetite suppressants (mainly fenfluramines and aminorex), human immunodeficiency virus (HIV) infection or portal hypertension, is indistinguishable from primary PAH, with regard to clinical course, histopathology and response to treatment. Therefore, a recent World Health Organization-sponsored consensus conference has suggested that the concept of primary PAH be extended to include these conditions and be renamed 'pulmonary arterial hypertension' (PAH).

At present, it remains unknown whether this concept of 'PAH' corresponds to a common pathogenic mechanism. Although understanding of the pathobiological mechanisms underlying IPAH has progressed rapidly over the past few years, it is still unfeasible to classify patients on a pathogenic basis and to define therapeutic approaches accordingly. Current treatments, including continuous infusion of prostaglandin (PG)I$_2$ and oral endothelin receptor antagonists, probably address downstream manifestations of the disease rather than the central pathogenic mechanisms. The identification of the FPAH gene, bone morphogenetic protein (BMP) receptor (BMPR)-II, and the recognition of central pathobiological abnormalities associated with IPAH, now provide a unique opportunity to develop a more robust understanding of the disease. In the near future, this should serve to assess new treatments aimed at correcting selective pulmonary vascular remodelling processes, and, simultaneously, to validate the pathophysiological concepts.

Identification of these molecular pathways might also provide insight into the understanding of secondary forms of PAH, including PAH secondary to chronic obstructive lung disease and left heart failure. In these conditions, as well as in persistent PAH in neonates, a genetic predisposition has been suggested. The severity of hypoxia-induced PAH also varies in intensity among individuals. Variations in expression and/or function of candidate genes involved in the process of pulmonary vascular remodelling might therefore improve understanding of secondary forms of PAH and also help to define susceptibility to PAH of various origins.

Pulmonary vascular remodelling

Pulmonary hypertension is associated with structural alterations in pulmonary arteries including intimal fibrosis, medial hypertrophy and adventitial changes [12].

Intimal lesions account for a great part of the reduction of luminal area of small arteries and potentially largely influence the overall pulmonary vascular resistance. Intimal lesions consist of eccentric intima thickening and fibrotic plexiform concentric and angiomatoid lesions. More advanced lesions acquire a fibrotic pattern, with interspersed myofibroblasts and marked accumulation of mucopolysaccharides. These lesions are widely present in explanted lungs of patients with severe PAH, both IPAH or PAH associated with CREST syndrome [13]. Variable degree of eccentric thickening can be seen also in cigarette smoker's lungs [14].

The increase in media thickness occurs by a combination of hypertrophy and hyperplasia of smooth muscle cells together with accumulation of extracellular matrix, including thicker but fragmented elastic laminæ. Fragmentation of elastic laminæ was initially described in PAH accompanying heart congenital defects [15] but is now recognised in all forms of PAH. In experimental PAH induced by the toxin monocrotaline, changes in elastin and collagen synthesis in the pulmonary artery walls, assessed both biochemically and ultrastructurally, were related to the development of progressive pulmonary hypertension with an increase in both insoluble elastin synthesis and total insoluble elastin content and an increase in collagen synthesis and total collagen content [16]. These changes have also been observed in other models of PAH, such as progressive pulmonary venous obstruction [17].

Noteworthy, the extracellular matrix is both a component of the thickened pulmonary vascular media and a regulator of smooth muscle cell growth. Tenascin, a glycoprotein involved in lung and vascular morphogenesis, which is strongly expressed in remodelled intima and medial layers of human hypertensive pulmonary arteries [18], is a potent regulator of smooth muscle cell proliferation [18–20]. Fibronectin changes smooth muscle cell phenotype from contractile to migratory.

The adventitia is mostly composed of fibroblasts. There is growing evidence that, rather than just a structural support to pulmonary vessels, the adventitia may also play a role in the regulation of pulmonary vascular function from the 'outside-in' (as comprehensively reviewed in [21]). The normal adventitia represents approximately 15% of the external diameter of pulmonary arteries larger than 50 μm in diameter. In IPAH arteries, the adventitial thickness increases to 28% of artery diameter, predominantly due to collagen deposition [22]. It also contains a perivascular cuff of inflammatory cells, which might modulate the growth of or transdifferentiate themselves into vascular structural cells in the pulmonary vascular wall.

Altogether these data show that extracellular matrix remodelling is a key component of PAH pathophysiology.

Proteases in experimental pulmonary hypertension

Matrix deposition is the result of increased matrix degradation insufficient to counterbalance excessive matrix synthesis [16, 23]. The normally tightly regulated degradation of extracellular matrix results from the activity of several proteases that are active at neutral pH and act in concert. Based on their active sites, two main classes of neutral proteases are the focus of interest: the serine-proteases, including the endogenous vascular elastase (EVE) and the plasminogen activator/plasmin system on one hand and the matrix metalloproteases (MMPs) also called the matrixins on the other hand.

EVE, the endogenous vascular elastase

Early studies, analysing the ultrastructure of pulmonary arteries on lung biopsy from patients with PAH, showed that the internal elastic lamina, which normally separates endothelial from smooth muscle cells in muscular arteries, is fragmented [15]. This suggested that an elastinolytic enzyme might be involved in the pathophysiology of the disease. This was further explored in experimental PAH induced in rats by the toxin monocrotalin, in which an increased number of breaks in the internal elastic lamina was associated with the initiation of vascular structural changes as early as 4 days after the toxin injection [24]. Subsequently, early increase in elastinolytic activity that precedes vascular changes and a later increase associated with progressive disease were confirmed, and the inhibition profile of enzymatic activity showed that it was attributable to a serine protease [23]. This increased elastinolysis was also shown in chronic hypoxia-induced experimental PAH [25]. Jacob et al. [26] and Hornebeck et al. [27] showed that a serine elastase was produced by aortic smooth muscle cells and associated with atherosclerosis, which was further characterised by Zhu et al. in pulmonary artery smooth muscle cells as the Endogenous Vascular Elastase (EVE) [28]. This elastase, like the poly-morphonuclear neutrophil serine elastase, is inhibited by α_1-antitrypsin/α_1-pro-teinase inhibitor (α_1-PI), α_2-macroglobulin, elafin and by some synthetic inhibitors [29], the most relevant inhibitor in vascular biology being elafin. EVE is a powerful enzyme that, by virtue of its ability to degrade elastin, will also degrade proteogly-cans that serve as storage sites for growth factors, such as basic fibroblast growth factor (FGF-2) and transforming growth factor. A study from Rabinovitch's group has shown that EVE releases FGF-2 in a biologically active form that stimulates smooth muscle cell proliferation [30]. Elastase activity either directly or *via* activa-tion of matrix metalloproteinases (see below) can induce production of the matrix glycoprotein tenascin (TN), which optimises the mitogenic response to FGF-2 and is, in fact, a prerequisite for the response to epidermal growth factor (EGF) in cul-tured smooth muscle cells [19, 20]. The process of smooth muscle cell migration also appears to depend, at least in experimental animals, on the continued activ-ity of elastase. Obviously, elastin peptides stimulate the production of the matrix glycoprotein fibronectin, which changes smooth muscle cells from a contractile to a migratory phenotype [31, 32].

The plasminogen system

Although these findings emphasise the important role of elastase in the pathogenesis of PAH, the question remains open regarding the role of other proteases specialised in extracellular matrix degradation, namely plasminogen activators/plasmin system and MMPs.

Several observations point to an additional role of the plasminogen system in pulmonary hypertension and to enhanced plasmin proteolysis that contributes to pulmonary vascular remodelling. The central reaction of the plasminogen activator (PA) system is the conversion of plasminogen to plasmin by plasminogen activators (PAs), the tissue plasminogen, tPA, and urokinase plasminogen activator, uPA [33]. The serine protease plasmin degrades fibrin to fibrin-degradation products. How-ever, plasmin has several substrates other than fibrin, including blood coagulation factors, cell surface receptors, some matrix metalloproteinases, that plasmin acti-vates, and structural components of the extracellular matrix such as fibronectin or laminin. While plasminogen resides primarily within the plasma, with the liver rep-resenting the primary site of plasminogen synthesis, plasminogen mRNA is present in several tissues, including adrenal, kidney, brain, testis, heart, lung, uterus, spleen, thymus, and gut, supporting a broadly distributed functional role of the PA system [34]. Tissue-type PA and u-PA activate plasminogen by cleaving a specific Arg-Val peptide bond located within the protease domain. The activation of plasminogen by t-PA is highly dependent on the presence of cofactors, such as fibrin, that bind and alter the conformation of plasminogen [35]. The activation of plasminogen by u-PA is tightly regulated at the cell surface, due to anchorage of u-PA to cell surface *via* a specific receptor, the uPA-R. Plasmin formation is intensely regulated by PA inhibitors, which inhibit t-PA and u-PA, most notably PA inhibitor-1 (PAI-1) [33]. Plasmin is directly inhibited by α_2-antiplasmin, which circulates in plasma.

In addition to functioning within the vascular lumen to control fibrinolysis, the PA system is active within the blood vessel wall, where it plays an important role in controlling vascular remodelling. The development of intimal hyperplasia after vascular injury is diminished in plasminogen-deficient mice, supporting the concept that plasmin associated with vascular smooth muscle cells enhances cell migration by fostering extracellular matrix degradation, either directly or indirectly by activating matrix metalloproteinases [36]. Vascular smooth muscle cells express u-PA and its receptor. Urokinase (u-PA) deficiency and pharmacological inhibition of u-PA receptor, but not t-PA deficiency, inhibit neointima formation in mice, suggesting that u-PA-triggered plasmin formation drives vascular smooth muscle cell migration [37–39]. Regarding pulmonary vascular system and pulmonary hypertension, hypoxia increases expression of u-PAR, enhances plasma fibrinolytic activity, and upregulates expression of plasminogen activators during ventricular hypertrophy in response to hypoxia or overloading [40–42]. Interestingly, and also similar to elastase, plasmin could also participate in smooth muscle cell proliferation indirectly as it has been shown to induce the release of FGF-2 from the extracellular matrix [43].

The matrix metalloproteases

The matrixins form a family of > 20 members known in humans, initially identified based on their ability to degrade extracellular matrix proteins (e.g., collagenases degrade fibrillar collagens, metallo-elastase, elastin, etc.), and are known to have many other roles [44]. One of the striking features of the matrixins is that many of those genes are 'inducible'. The effectors include growth factors, cytokines, chemical agents (e.g., phorbol esters, actin stress fibre-disrupting drugs), physical stress and oncogenic cellular transformation. MMP gene expression may be downregulated by suppressive factors (e.g., transforming growth factor, retinoic acids, glucocorticoids). Their proteolytic activities are precisely controlled during activation from their precursors and inhibition by endogenous inhibitors, α-macroglobulins and tissue inhibitors of metalloproteinases (TIMPs). All matrixins are synthesised as pre-pro-enzymes, the loss of the 'pre' peptide is a signal of secretion. Apart from a few members activated intracellularly by furin, most are secreted from the cell as inactive zymogens. Secreted promatrixins are activated *in vitro* by proteinases, such as plamin, and by nonproteolytic agents, such as *in vivo* reactive oxygen species and *in vitro*, thiol-reactive agents, mercurial compounds, and denaturants. In all cases, activation requires the disruption of the Cys-Zn^{2+} (cystein switch) interaction between the zinc of the active site and the cystein present within the 'pro'-domain. Subsequently, the removal of the propeptide proceeds often in a step-wise manner [45]. *In vivo*, most promatrixins are likely to be activated by tissue or plasma proteinases, such as neutrophil elastase [46] or plasmin, or opportunistic bacterial proteinases. Using transgenic mice deficient in uPA, Carmeliet et al. have shown that the uPA/plasmin system is, *in vivo*, a pathophysiologically significant activator of promatrixins [45]. By contrast, no data exist regarding activation of promatrixin by EVE. Other activation cascades are known; for example, the activation of pro-MMP-2 takes place primarily on the cell surface and results from the action of membrane-anchored MMPs, the 'membrane-type MMPs' (MT-MMPs). Studies have shown that this activation process requires both active MT1-MMP and the TIMP-2-bound MT1-MMP [47].

TIMPs (21–30 kDa) are the major endogenous inhibitors of MMP activities in tissue, and four homologous TIMPs (TIMPs 1–4) have been identified to date [48]. TIMPs exhibit additional biological functions. As detailed above, TIMP-2 plays a role in MMP-2 activation. TIMP-1 and TIMP-2 have mitogenic activities on a number of cell types, whereas overexpression of these inhibitors reduces tumour cell growth, and TIMP-2, but not TIMP-1, inhibits FGF-2 induced human endothelial cell growth. These biological activities of TIMPs are independent of MMP-inhibitory activities [49].

MMPs, particularly gelatinase A/MMP-2, which degrade the type IV collagen of basement membranes, are increased in the pulmonary vascular bed, during both toxin- and hypoxia-induced experimental PAH [50]. Increases in interstitial collagenase (MMP-13), stromelysin-1 (MMP-3) and gelatinases A (MMP-2) and B (MMP-9) have also been described following return to normoxia [51, 52]. Interestingly, inhibition of MMPs by either a synthetic inhibitor, doxycycline, or adenovirus-mediated human TIMP-1 gene transfer during chronic hypoxia is associated with exacerbation of PAH and vascular remodelling. Either of two MMP-inhibiting treatments increased muscularisation and collagen accumulation in small pulmonary arteries [53], providing strong support for the argument that MMPs play a crucial protective role in hypoxic PAH. In keeping with these results is the demonstration that deficiency of uPA-mediated plasmin generation impairs vascular remodelling in hypoxic PAH [54]. MMPs and plasmin probably protect against PAH by limiting matrix deposition. Another important hypothesis concerns angiogenesis, which represents an important protective mechanism, as demonstrated by increase in lung vascular endothelial growth factor (VEGF) during hypoxic PAH, improvement of PAH after VEGF gene transfer and worsening of PAH following angiostatin gene transfer [55, 56].

In contrast to hypoxic PAH, inhibition of MMPs in an *ex vivo* model of organ culture of pulmonary artery rings obtained from rats treated by monocrotalin, showed regression of medial thickness to control levels [57]. Similar results in this study, obtained by either serine-protease inhibitors or metalloprotease inhibitors, come from tight interactions between the two proteolytic systems, as elastase can activate some pro-MMPs [46] and degrade the MMP-inhibitor, TIMP-1 [57]. In contrast, most MMPs degrade the major elastase inbihitor, α_1-PI [46]. In keeping with the *ex vivo* results, *in vivo* experiments in toxic monocrotaline-induced PAH have shown that inhibition of MMPs by adenovirus-mediated human TIMP-1 gene transfer in the lung leads to less severe PAH with decreased muscularisation and increased lung-cell apoptosis, as compared to controls [58]. The effect of TIMP-1 on PAH is consistent with an ability of MMP inhibition to prevent monocrotaline-induced pulmonary vascular remodelling and PAH, in part by reducing smooth muscle cell migration and proliferation. All together, these data support a synergistic and deleterious role for serine- and metalloproteases in toxic PAH and indicate that MMPs may have opposite effects in different PAH models.

Proteases and BMPs

BMPs are the largest group of cytokines within the TGF-β superfamily and were originally identified as molecules regulating growth and differentiation of bone and cartilage [59]. BMPs regulate growth, differentiation, and apoptosis in a diverse number of cell lines, including mesenchymal and epithelial cells, acting as instructive signals during embryogenesis and contributing to the maintenance and repair of adult tissues [59, 60]. TGF-β superfamily type II receptors are constitutively active serine-threonine kinases and form homodimers that exist constitutively or are recruited to receptor complexes on ligand stimulation. BMPR-II is distinguished from other TGF-β superfamily type II receptors by a long carboxyl-terminal sequence following the intracellular kinase domain [61]. BMPR-II initiates intracellular signalling in response to specific ligands: BMP 2, BMP 4, BMP 6, BMP 7, growth and differentiation factor-5 (GDF 5), and GDF 6 [61].

Mutations of *BMPR-II* have been found in FPAH [10]. Approximately 30% of mutations are missense mutations occurring in highly conserved amino acids with predictable effects on receptor function. For example, many of these involve the serine-threonine kinase domain of BMPR-II or the extracellular ligand binding domain. However, the majority (~70%) of *BMPR2* coding mutations are frameshift and nonsense mutations, many of which would be expected to produce a transcript susceptible to nonsense-mediated mRNA decay. Thus, haploinsufficiency for BMPR-II represents the predominant molecular mechanism underlying inherited predisposition to FPAH. Further genetic analysis is revealing an increasing number of families in which BMPR-II mutation is implicated, including the identification of gene deletions and rearrangements. Interestingly, BMPR-II alterations are also involved in experimental PAH, either induced by chronic hypoxia [62] or by the toxic model, monocrotalin-induced [63].

Clearly, one major mechanism by which BMPR-II plays a role in FPAH is smooth muscle proliferation. However, interference with extracellular matrix remodelling can be hypothesised. Very recently, induction of uPA upon BMP 4 stimulation of BMPR-II has been reported in tumoral cell line [64]. This report is the only one to date that has explored induction of protease(s) after engagement of the BMPr-II receptor but is certainly a promising direction.

MMPs in human pulmonary hypertension

Very few studies are available in the human disease, regarding MMPs. Based on studies using cultured human smooth muscle cells isolated from elastic pulmonary arteries from IPAH patients obtained surgically, we reported MMP/TIMP production by smooth muscle cells *in vitro*. We also performed immunolocalisation in whole medium-sized pulmonary arteries. *In vitro*, TIMP-1 was overexpressed and MMP-3 underexpressed by IPAH cells, whereas MMP-1 expression was similar in the two groups. Total MMP-2 overexpression was also found, with a greater amount of the active form in IPAH cells as compared with controls. *In situ* studies showed gelatinolytic activity in tissue sections and strong MMP-2 immunostaining along the inner elastic lamina up to the lamina break. TIMP-1 immunostaining was found in both control and IPAH arteries, whereas MMP-3 staining was detected only in the media of a few control specimens [65]. It is unclear whether a similar pattern is present in distally remodelled pulmonary arteries; however, endothelial cells express moderate/intense immunohistochemical expression of MMP-2, while myofibroblasts display low levels of this extracellular protease [66]. Membrane type-1-MMP was also expressed in endothelial and myofibroblastic cells of concentric and plexiform lesions.

Demonstration of a TIMP-1–MMP imbalance conducive to extracellular matrix accumulation does not rule out a role for active MMP-2 in IPAH. Proteolysis may be effective in limited areas even when TIMP levels are high in the extracellular milieu, because MMP-2 tethering and activation at the cell surface focuses the catalytic activity on limited targets on the cell membrane. This hypothesis is supported by immunohistology and *in situ* zymography data, which clearly show that gelatinolytic activity colocalised with MMP-2 immunostaining in arteries from IPAH patients [65]. This pattern of MMP and TIMP expression, characterised by increased TIMP-1 levels coexisting with evidence of extracellular matrix degradation, has been found in other fibrotic diseases, such as adult respiratory distress syndrome (ARDS). In bronchoalveolar lavage (BAL) fluid from ARDS patients, TIMP-1 levels were significantly higher than in healthy controls. Despite the high TIMP-1 levels, extracellular matrix degradation by MMPs is suggested by the presence of active MMP-2 in epithelial lining fluid [67] and of basement membrane disruption markers in BAL fluid of ARDS patients [68]. Altogether, these data suggest that a TIMP-1–MMP imbalance promoting extracellular matrix accumulation within the interstitial tissue may coexist with the presence of active MMP-2 confined to the cell surface.

In IPAH, disruption of the internal elastic lamina, extracellular matrix disorganisation and smooth muscle cell migration are strong arguments supporting a direct role for active MMP-2. This enzyme not only degrades nonfibrillar collagen, but also cleaves elastin. Moreover, latent-MMP-2 may both bind to elastin and undergo auto-activation, subsequently degrading elastin [69]. These results on MMP-2 expression in PAH are consistent with previous data, as *in situ* zymography and MMP-2 immunolocalisation showed colocalisation of gelatinolytic activities and MMP-2 along elastic fibres. Also, active MMP-2 may contribute not only to extracellular matrix remodelling but also to important processes in IPAH, such as smooth muscle cell migration and proliferation [70].

Inflammation and pulmonary hypertension

There is compelling evidence of global immunological alterations in IPAH patients [71] and PAH occurs in the setting of profound immune deregulation underlying HIV infection and collagen vascular diseases. The recognition of an inflammatory component in PAH [12] supports the investigation of expression of cytokines that might potentially drive perivascular inflammation and thus contribute to the disease. Remodelled pulmonary arteries express IL-1, IL-6, and PDGF in infiltrating inflammatory cells [78, 79], the chemokine RANTES (acronym for regulated upon activation, normal T cell expressed and secreted), an important chemoattractant for monocytes and T cells, and the macrophage inflammatory protein-1α (MIP-1α). Lungs of IPAH patients have increased expression of fraktaline, a chemokine involved in T cell trafficking and monocyte recruitment, and their circulating CD4 and CD8 T cells have higher levels of the fraktaline receptor CX3CR1 when compared with controls or samples of patients with thromboembolic PAH.

Inflammatory cells infiltrating remodelled pulmonary arteries may include subpopulations of vascular precursor or early-progenitor cells, also potential contributors to pulmonary vascular remodelling in PAH. Pulmonary arteries in PAH caused by chronic hypoxia contain an infiltrating subpopulation of fibrocytes, identified by the expression mononuclear cell markers CD45, CD11b, CD14, and the fibroblast marker α1-procollagen. About 15% and 20% of these cells also undergo proliferation and express smooth muscle α-actin, respectively [72, 73]. These studies also document that depletion of circulating monocytic cells alleviates pulmonary vascular remodelling caused by chronic hypoxia. Endothelial cell precursors may play a beneficial role in PAH since their administration to monocrotaline-treated rats has dramatic healing effects in remodelled pulmonary arteries, notably when transfected with the endothelial nitric oxide synthase gene [74]. Pulmonary vascular inflammation has also been documented in chronic obstructive pulmonary disease with or without coexisting PAH [14, 75–78], and vascular progenitor cells identified in lung vessels [77].

Within the scope of the present review, it should be stressed that protease production is part of the inflammatory response. Sources of protease are circulating cells, leukocytes [79, 80] and platelets [81, 82], together with resident endothelial [83] and smooth muscle cells [84].

Therapeutic consequences and future directions

In experimental models, protease inhibition, especially elastase inhibition has clearly proven its efficacy, both at preventing and curing PAH induced by monocrotalin [18, 57]. Accordingly, overexpression of the EVE inhibitor elafin protects partially transgenic mice from hypoxic PAH [85]. However, conflicting results have been obtained regarding metalloprotase inhibition, depending on the considered model: worsening hypoxic PAH [53] and preventing toxic monocrotalin-induced PAH [58].

It is certainly worth asking the question "should we expect an improvement in PAH when using protease inhibitors?". It seems quite clear-cut when thinking of elastic lamina fragmentation, but not so obvious when taking into account the large increase in extracellular matrix deposition observed in pulmonary vascular beds and regarding all other roles for these proteases in different cellular processes, such as angiogenesis, cell migration and cell differentiation. All these considerations leave the question of their beneficial or deleterious role open. The tight interplay between the three proteolytic systems further complicates the answer.

References

1   Farber HW, Loscalzo J (2004) Pulmonary arterial hypertension. *N Engl J Med* 351: 1655–1665
2   Simonneau G, Galie N, Rubin LJ, Langleben D, Seeger W, Domenighetti G, Gibbs S, Lebrec D, Speich R, Beghetti M et al (2004) Clinical classification of pulmonary hypertension. *J Am Coll Cardiol* 43: 5S–12S
3   Gaine SP, Rubin LJ (1998) Primary pulmonary hypertension. *Lancet* 352: 719–725
4   Hoeper MM, Galie N, Simonneau G, Rubin LJ (2002) New treatments for pulmonary arterial hypertension. *Am J Respir Crit Care Med* 165: 1209–1216
5   Abenhaim L, Moride Y, Brenot F, Rich S, Benichou J, Kurz X, Higenbottam T, Oakley C, Wouters E, Aubier M et al (1996) Appetite-suppressant drugs and the risk of primary

pulmonary hypertension. International Primary Pulmonary Hypertension Study Group. *N Engl J Med* 335: 609–616

6   Deng Z, Haghighi F, Helleby L, Vanterpool K, Horn EM, Barst RJ, Hodge SE, Morse JH, Knowles JA (2000) Fine mapping of PPH1, a gene for familial primary pulmonary hypertension, to a 3-cM region on chromosome 2q33. *Am J Respir Crit Care Med* 161: 1055–1059

7   Deng Z, Morse JH, Slager SL, Cuervo N, Moore KJ, Venetos G, Kalachikov S, Cayanis E, Fischer SG, Barst RJ et al (2000) Familial primary pulmonary hypertension (gene PPH1) is caused by mutations in the bone morphogenetic protein receptor-II gene. *Am J Hum Genet* 67: 737–744

8   Lane KB, Machado RD, Pauciulo MW, Thomson JR, Phillips JA 3rd, Loyd JE, Nichols WC, Trembath RC (2000) Heterozygous germline mutations in BMPR2, encoding a TGF-beta receptor, cause familial primary pulmonary hypertension. The International PPH Consortium. *Nat Genet* 26: 81–84

9   Nichols WC, Koller DL, Slovis B, Foroud T, Terry VH, Arnold ND, Siemieniak DR, Wheeler L, Phillips JA 3rd, Newman JH et al (1997) Localization of the gene for familial primary pulmonary hypertension to chromosome 2q31-32. *Nat Genet* 15: 277–280

10   Machado RD, Aldred MA, James V, Harrison RE, Patel B, Schwalbe EC, Gruenig E, Janssen B, Koehler R, Seeger W et al (2006) Mutations of the TGF-beta type II receptor BMPR2 in pulmonary arterial hypertension. *Hum Mutat* 27: 121–132

11   Thomson JR, Machado RD, Pauciulo MW, Morgan NV, Humbert M, Elliott GC, Ward K, Yacoub M, Mikhail G, Rogers P et al (2000) Sporadic primary pulmonary hypertension is associated with germline mutations of the gene encoding BMPR-II, a receptor member of the TGF-beta family. *J Med Genet* 37: 741–745

12   Tuder RM, Marecki JC, Richter A, Fijalkowska I, Flores S (2007) Pathology of pulmonary hypertension. *Clin Chest Med* 28: 23–42, vii

13   Cool CD, Kennedy D, Voelkel NF, Tuder RM (1997) Pathogenesis and evolution of plexiform lesions in pulmonary hypertension associated with scleroderma and human immunodeficiency virus infection. *Hum Pathol* 28: 434–442

14   Santos S, Peinado VI, Ramirez J, Melgosa T, Roca J, Rodriguez-Roisin R, Barbera JA (2002) Characterization of pulmonary vascular remodelling in smokers and patients with mild COPD. *Eur Respir J* 19: 632–638

15   Rabinovitch M, Bothwell T, Hayakawa BN, Williams WG, Trusler GA, Rowe RD, Olley PM, Cutz E (1986) Pulmonary artery endothelial abnormalities in patients with congenital heart defects and pulmonary hypertension. A correlation of light with scanning electron microscopy and transmission electron microscopy. *Lab Invest* 55: 632–653

16   Todorovitch-Hunter L, Johnson DJ, Ranger P, Keeley FW, Rabinovitsh M (1988) Altered elastin and collagen synthesis associated with progressive pulmonary hypertension induced by monocrotaline: a biochemical and ultrastructural study. *Lab Invest* 58: 184–195

17   LaBourene JI, Coles JG, Johnson DJ, Mehra A, Keeley FW, Rabinovitch M (1990) Alterations in elastin and collagen related to the mechanism of progressive pulmonary

venous obstruction in a piglet model. A hemodynamic, ultrastructural, and biochemical study. *Circ Res* 66: 438–456

18 Cowan K, Jones P, Rabinovitch M (2000) Elastase and matrix metalloproteinase inhibitors induce regression, and tenascin-C antisense prevents progression, of vascular disease. *J Clin Invest* 105: 21–34

19 Jones P, Cowan K, Rabinovitch M (1997) Tenascin-C, proliferation and subendothelial accumulation of fibronectin in progressive pulmonary vascular disease. *Am J Pathol* 150: 1349–1360

20 Jones P, Crack J, Rabinovitch M (1997) Regulation of Tenascin-C, a vascular smooth muscle cell survival factor that interacts with the alphaVbeta3 integrin to promote EGF receptor phosphorylation and growth. *J Cell Biol* 139: 279–293

21 Stenmark KR, Davie N, Frid M, Gerasimovskaya E, Das M (2006) Role of the adventitia in pulmonary vascular remodeling. *Physiology (Bethesda)* 21: 134–145

22 Chazova I, Loyd JE, Zhdanov VS, Newman JH, Belenkov Y, Meyrick B (1995) Pulmonary artery adventitial changes and venous involvement in primary pulmonary hypertension. *Am J Pathol* 146: 389–397

23 Todorovitch-Hunter L, Dodo H, Ye C, McCready L, Keeley FW, Rabinovitch M (1992) Increased pulmonary artery elastolytic activity in adult rats with monocrotaline-induced progressive hypertensive pulmonary vascular disease compared with infant rats with non-progressive disease. *Am Rev Respir Dis* 146: 213–233

24 Ye CL, Rabinovitch M (1991) Inhibition of elastolysis by SC-37698 reduces development and progression of monocrotaline pulmonary hypertension. *Am J Physiol* 261: H1255–1267

25 Maruyama K, Ye CL, Woo M, Venkatacharya H, Lines LD, Silver MM, Rabinovitch M (1991) Chronic hypoxic pulmonary hypertension in rats and increased elastolytic activity. *Am J Physiol* 261: H1716–1726

26 Jacob MP, Bellon G, Robert L, Hornebeck W, Ayrault-Jarrier M, Burdin J, Polonovski J (1981) Elastase-type activity associated with high density lipoproteins in human serum. *Biochem Biophys Res Commun* 103: 311–318

27 Hornebeck W, Derouette JC, Robert L (1975) Isolation, purification and properties of aortic elastase. *FEBS Lett* 58: 66–70

28 Zhu L, Wigle D, Hinek A, Kobayashi J, Ye C, Zuker M, Dodo H, Keeley FW, Rabinovitch M (1994) The endogenous vascular elastase that governs development and progression of monocrotaline-induced pulmonary hypertension in rats is a novel enzyme related to the serine proteinase adipsin. *J Clin Invest* 94: 1163–1171

29 Rabinovitch M (1999) EVE and beyond, retro and prospective insights. *Am J Physiol* 277: L5–12

30 Thompson K, Rabinovitch M (1996) Exogenous leukocyte and endogenous elastases can mediate mitogenic activity in pulmonary artery smooth muscle cells by release of extracellular-matrix bound basic fibroblast growth factor. *J Cell Physiol* 166: 495–505

31 Hinek A, Boyle J, Rabinovitch M (1992) Vascular smooth muscle cell detachment from elastin and migration through elastic laminae is promoted by chondroitin sulfate-

induced „shedding" of the 67-kDa cell surface elastin binding protein. *Exp Cell Res* 203: 344–353

32 Hinek A, Molossi S, Rabinovitch M (1996) Functional interplay between interleukin-1 receptor and elastin binding protein regulates fibronectin production in coronary artery smooth muscle cells. *Exp Cell Res* 225: 122–131

33 Fay WP, Garg N, Sunkar M (2007) Vascular functions of the plasminogen activation system. *Arterioscler Thromb Vasc Biol* 27: 1231–1237

34 Zhang L, Seiffert D, Fowler BJ, Jenkins GR, Thinnes TC, Loskutoff DJ, Parmer RJ, Miles LA (2002) Plasminogen has a broad extrahepatic distribution. *Thromb Haemost* 87: 493–501

35 Kolev K, Machovich R (2003) Molecular and cellular modulation of fibrinolysis. *Thromb Haemost* 89: 610–621

36 Carmeliet P, Moons L, Ploplis V, Plow E, Collen D (1997) Impaired arterial neointima formation in mice with disruption of the plasminogen gene. *J Clin Invest* 99: 200–208

37 Carmeliet P, Moons L, Herbert JM, Crawley J, Lupu F, Lijnen R, Collen D (1997) Urokinase but not tissue plasminogen activator mediates arterial neointima formation in mice. *Circ Res* 81: 829–839

38 Quax PH, Lamfers ML, Lardenoye JH, Grimbergen JM, de Vries MR, Slomp J, de Ruiter MC, Kockx MM, Verheijen JH, van Hinsbergh VW (2001) Adenoviral expression of a urokinase receptor-targeted protease inhibitor inhibits neointima formation in murine and human blood vessels. *Circulation* 103: 562–569

39 Schafer K, Konstantinides S, Riedel C, Thinnes T, Muller K, Dellas C, Hasenfuss G, Loskutoff DJ (2002) Different mechanisms of increased luminal stenosis after arterial injury in mice deficient for urokinase- or tissue-type plasminogen activator. *Circulation* 106: 1847–1852

40 Bansal DD, Klein MR, Hausmann EHS, MacGregor RR (1997) Secretion of cardiac plasminogen activator during hypoxia-induced right ventricular hypertrophy. *J Mol Cell Cardiol* 29: 310563114

41 Graham CH, Fitzpatrick TE, McCrae KR (1998) Hypoxia stimulates urokinase receptor expression through a heme protein-dependent pathway. *Blood* 91: 3300–3307

42 Pinsky DJ, Liao H, Lawson CA, Yan SF, Chen J, Carmeliet P, Loskutoff DJ, Stern DM (1998) Coordinated induction of plasminogen activator inhibitor-1 (PAI-1) and inhibition of plasminogen activator gene expression by hypoxia promotes pulmonary vascular fibrin deposition. *J Clin Invest* 102: 919–928

43 Saksela O, Rifkin DB (1990) Release of basic fibroblast growth factor-heparan sulfate complexes from endothelial cells by plasminogen activator-mediated proteolytic activity. *J Cell Biol* 110: 767–775

44 Overall CM, Lopez-Otin C (2002) Strategies for MMP inhibition in cancer: innovations for the post-trial era. *Nat Rev Cancer* 2: 657–672

45 Nagase H (1997) Activation mechanisms of matrix metalloproteinases. *Biol Chem* 378: 151–160

46 Okada Y, Nakanishi I (1989) Activation of matrix metalloproteinase 3 (stromelysin)

and matrix metalloproteinase 2 ('gelatinase') by human neutrophil elastase and cathepsin G. *FEBS Lett* 249: 353–356

47 Butler GS, Butler MJ, Atkinson SJ, Will H, Tamura T, Schade van Westrum S, Crabbe T, Clements J, d'Ortho M-P, Murphy G (1998) The TIMP-2-MT1 MMP 'receptor' regulates the concentration and efficient activation of progelatinase A. A kinetic study. *J Biol Chem* 273: 871–880

48 Brew K, Dinakarpandian D, Nagase H (2000) Tissue inhibitors of metalloproteinases: evolution, structure and function. *Biochim Biophys Acta* 1477: 267–283

49 Hoegy SE, Oh HR, Corcoran ML, Stetler-Stevenson WG (2001) Tissue inhibitor of metalloproteinases-2 (TIMP-2) suppresses TKR-growth factor signaling independent of metalloproteinase inhibition. *J Biol Chem* 276: 3203–3214

50 Frisdal E, Gest V, Vieillard-Baron A, Levame M, Lepetit H, Eddahibi S, Lafuma C, Harf A, Adnot S, Dortho MP (2001) Gelatinase expression in pulmonary arteries during experimental pulmonary hypertension. *Eur Respir J* 18: 838–845

51 Thakker-Varia S, Tozzi CA, Poiani GJ, Barbiaz JP, Tatem L, Wilson FJ, Riley DJ (1998) Expression of matrix degrading enzymes in pulmonary vascular remodeling in the rat. *Am J Physiol* (*Lung Cell Mol Physiol* 19) 275: L398–L406

52 Tozzi CA, Thakker-Varia S, Shiu YY, Bannett RF, Peng BW, Poiani GJ, Wilson FJ, Riley DJ (1998) Mast cell colagenase correlates with regression of pulmonary vascular remodeling in the rat. *Am J Respir Cell Mol Biol* 18: 497–510

53 Vieillard-Baron A, Frisdal E, Eddahibi S, Deprez I, Baker AH, Newby AC, Berger P, Levame M, Raffestin B, Adnot S et al (2000) Inhibition of matrix metalloproteinases by lung TIMP-1 gene transfer or doxycycline aggravates pulmonary hypertension in rats. *Circ Res* 87: 418–425

54 Levi M, Moons L, Bouche A, Shapiro SD, Collen D, Carmeliet P (2001) Deficiency of urokinase-type plasminogen activator-mediated plasmin generation impairs vascular remodeling during hypoxia-induced pulmonary hypertension in mice. *Circulation* 103: 2014–2020

55 Partovian C, Adnot S, Eddahibi S, Teiger E, Levame M, Dreyfus P, Raffestin B, Frelin C (1998) Heart and lung VEGF mRNA expression in rats with monocrotaline- or hypoxia-induced pulmonary hypertension. *Am J Physiol* 275: H1948–1956

56 Partovian C, Adnot S, Raffestin B, Louzier V, Levame M, Mavier IM, Lemarchand P, Eddahibi S (2000) Adenovirus-mediated lung vascular endothelial growth factor overexpression protects against hypoxic pulmonary hypertension in rats. *Am J Respir Cell Mol Biol* 23: 762–771

57 Cowan K, Heilbut A, Humpl T, Lam C, Ito S, Rabinovitch M (2000) Complete reversal of fatal pulmonary hypertension in rats by a serine elastase inhibitor. *Nat Med* 6: 698–702

58 Vieillard-Baron A, Frisdal E, Raffestin B, Baker AH, Eddahibi S, Adnot S, D'Ortho MP (2003) Inhibition of matrix metalloproteinases by lung TIMP-1 gene transfer limits monocrotaline-induced pulmonary vascular remodeling in rats. *Hum Gene Ther* 14: 861–869

59 Miyazono K, Maeda S, Imamura T (2005) BMP receptor signaling: transcriptional targets, regulation of signals, and signaling cross-talk. *Cytokine Growth Factor Rev* 16: 251–263

60 Massague J, Chen YG (2000) Controlling TGF-beta signaling. *Genes Dev* 14: 627–644

61 Rosenzweig BL, Imamura T, Okadome T, Cox GN, Yamashita H, ten Dijke P, Heldin CH, Miyazono K (1995) Cloning and characterization of a human type II receptor for bone morphogenetic proteins. *Proc Natl Acad Sci USA* 92: 7632–7636

62 Takahashi H, Goto N, Kojima Y, Tsuda Y, Morio Y, Muramatsu M, Fukuchi Y (2006) Downregulation of type II bone morphogenetic protein receptor in hypoxic pulmonary hypertension. *Am J Physiol Lung Cell Mol Physiol* 290: L450–458

63 Morty RE, Nejman B, Kwapiszewska G, Hecker M, Zakrzewicz A, Kouri FM, Peters DM, Dumitrascu R, Seeger W, Knaus P et al (2007) Dysregulated bone morphogenetic protein signaling in monocrotaline-induced pulmonary arterial hypertension. *Arterioscler Thromb Vasc Biol* 27: 1072–1078

64 Deng H, Makizumi R, Ravikumar TS, Dong H, Yang W, Yang WL (2007) Bone morphogenetic protein-4 is overexpressed in colonic adenocarcinomas and promotes migration and invasion of HCT116 cells. *Exp Cell Res* 313: 1033–1044

65 Lepetit H, Eddahibi S, Fadel E, Frisdal E, Munaut C, Noel A, Humbert M, Adnot S, D'Ortho MP, Lafuma C (2005) Smooth muscle cell matrix metalloproteinases in idiopathic pulmonary arterial hypertension. *Eur Respir J* 25: 834–842

66 Matsui K, Takano Y, Yu ZX, Hi JE, Stetler-Stevenson WG, Travis WD, Ferrans VJ (2002) Immunohistochemical study of endothelin-1 and matrix metalloproteinases in plexogenic pulmonary arteriopathy. *Pathol Res Pract* 198: 403–412

67 Delclaux C, d'Ortho M-P, Delacourt C, Lebargy F, Brun-Buisson C, Brochard L, Lemaire F, Lafuma C, Harf A (1997) Gelatinase in epithelial lining fluid of patients with adult respiratory distress syndrome. *Am J Physiol (Lung Cell Mol Physiol 16)* 272: L442–L451

68 Kondoh Y, Tanagushi H, Taki F, Takagi K, Satake T (1992) 7S collagen in bronchoalveolar lavage fluid of patients with adult respiratory distress syndrome. *Chest* 101: 1091–1094

69 Emonard H, Hornebeck W (1997) Binding of 92 kDa and 72 kDa progelatinases to insoluble elastin modulates their proteolytic activation. *Biol Chem* 378: 265–271

70 Uzui H, Lee JD, Shimizu H, Tsutani H, Ueda T (2000) The role of protein-tyrosine phosphorylation and gelatinase production in the migration and proliferation of smooth muscle cells. *Atherosclerosis* 149: 51–59

71 Dorfmuller P, Perros F, Balabanian K, Humbert M (2003) Inflammation in pulmonary arterial hypertension. *Eur Respir J* 22: 358–363

72 Frid MG, Brunetti JA, Burke DL, Carpenter TC, Davie NJ, Reeves JT, Roedersheimer MT, van Rooijen N, Stenmark KR (2006) Hypoxia-induced pulmonary vascular remodeling requires recruitment of circulating mesenchymal precursors of a monocyte/macrophage lineage. *Am J Pathol* 168: 659–669

73 Frid MG, Brunetti JA, Burke DL, Carpenter TC, Davie NJ, Stenmark KR (2005) Circulating mononuclear cells with a dual, macrophage-fibroblast phenotype contribute robustly to hypoxia-induced pulmonary adventitial remodeling. *Chest* 128: 583S–584S

74 Zhao YD, Courtman DW, Deng Y, Kugathasan L, Zhang Q, Stewart DJ (2005) Rescue of monocrotaline-induced pulmonary arterial hypertension using bone marrow-derived endothelial-like progenitor cells: efficacy of combined cell and eNOS gene therapy in established disease. *Circ Res* 96: 442–450

75 Barbera JA, Peinado VI, Santos S (2003) Pulmonary hypertension in chronic obstructive pulmonary disease. *Eur Respir J* 21: 892–905

76 Peinado VI, Barbera JA, Abate P, Ramirez J, Roca J, Santos S, Rodriguez-Roisin R (1999) Inflammatory reaction in pulmonary muscular arteries of patients with mild chronic obstructive pulmonary disease. *Am J Respir Crit Care Med* 159: 1605–1611

77 Peinado VI, Ramirez J, Roca J, Rodriguez-Roisin R, Barbera JA (2006) Identification of vascular progenitor cells in pulmonary arteries of patients with chronic obstructive pulmonary disease. *Am J Respir Cell Mol Biol* 34: 257–263

78 Santos S, Peinado VI, Ramirez J, Morales-Blanhir J, Bastos R, Roca J, Rodriguez-Roisin R, Barbera JA (2003) Enhanced expression of vascular endothelial growth factor in pulmonary arteries of smokers and patients with moderate chronic obstructive pulmonary disease. *Am J Respir Crit Care Med* 167: 1250–1256

79 Shapiro S (1994) Elastolytic metalloproteinases produced by human mononuclear phagocytes. Potential roles in destructive lung disease. *Am J Respir Crit Care Med* 150: S160–164

80 Weiss SJ, Peppin GJ (1986) Collagenolytic metalloenzymes of the human neutrophil: characteristics, regulation and potential function *in vivo. Biochem Pharmacol* 35: 3189–3197

81 Kazes I, Elalamy I, Sraer JD, Hatmi M, Nguyen G (2000) Platelet release of trimolecular complex components MT1-MMP/TIMP2/MMP2: involvement in MMP2 activation and platelet aggregation. *Blood* 96: 3064–3069

82 Sawicki G, Salas E, Murat J, Miszta Lane H, Radomski MW (1997) Release of gelatinase A during platelet activation mediates aggregation. *Nature* 386: 616–619

83 Koolwijk P, Sidenius N, Peters E, Sier CF, Hanemaaijer R, Blasi F, van Hinsbergh VW (2001) Proteolysis of the urokinase-type plasminogen activator receptor by metalloproteinase-12: implication for angiogenesis in fibrin matrices. *Blood* 97: 3123–3131

84 Bendeck MP, Zempo N, Clowes AW, Galardy RE, Reidy MA (1994) Smooth muscle cell migration and matrix metalloproteinase expression after arterial injury in the rat. *Circ Res* 75: 539–545

85 Zaidi SH, You XM, Ciura S, Husain M, Rabinovitch M (2002) Overexpression of the serine elastase inhibitor elafin protects transgenic mice from hypoxic pulmonary hypertension. *Circulation* 105: 516–521

MMP roles in the initiation and progression of cardiac remodeling leading to congestive heart failure

Jing Lin and Merry L. Lindsey

Medicine/Cardiology, University of Texas Health Science Center, 7703 Floyd Curl Drive, Mail Code 7872, San Antonio, TX 78229-3900, USA

Abstract

Cardiac remodeling is induced by a multitude of stimuli, including myocardial infarction, pressure and/or volume overload, and genetics; and involves matrix metalloproteinases (MMPs) every step of the way [1]. While cardiac remodeling is initially a compensatory response, the transition to adverse remodeling frequently culminates in the development of congestive heart failure (CHF), and CHF is a significant contributor to cardiovascular morbidity and mortality rates. This chapter will define cardiac remodeling, describe MMP-dependent mechanisms that stimulate the remodeling process, and explore future directions and therapeutic potentials in terms of MMP inhibition. We will focus on the primary diseases that stimulate cardiac remodeling in humans, namely myocardial infarction and hypertension. Understanding how cardiac remodeling evolves from an initially beneficial mechanism to a maladaptive event and how MMPs influence these stages will potentially provide us with markers to predict adverse events in the clinic.

Definitions of cardiac remodeling and congestive heart failure

First coined by Dr. Janice Pfeffer in the mid 1980s [2, 3], the term cardiac remodeling initially described the left ventricular response to myocardial infarction and included wall thinning, dilation, and infarct expansion; inflammation and necrotic myocyte resorption; and fibroblast accumulation and scar formation [4–6]. We have since included hypertension as a separate mechanism that induces cardiac remodeling, albeit through slightly different mechanisms. Cardiac remodeling is influenced by changes in inflammatory response (neutrophil and macrophage influx), hemodynamic load, molecular changes (neurohormonal activation and cytokine production), and extracellular responses (fibrosis and activation of matrix metalloproteinases) [7]. In response to a myriad of stimuli, molecular, cellular, and extracellular mechanisms combine to alter wall structure, chamber geometry, and LV function [8].

Cardiac remodeling continually cycles through several phases (Fig. 1) [9]. First is the injury response, where collagen is degraded and MMPs are induced to stimulate

Matrix Metalloproteinases in Tissue Remodelling and Inflammation,
edited by Vincent Lagente and Elisabeth Boichot
© 2008 Birkhäuser Verlag Basel/Switzerland

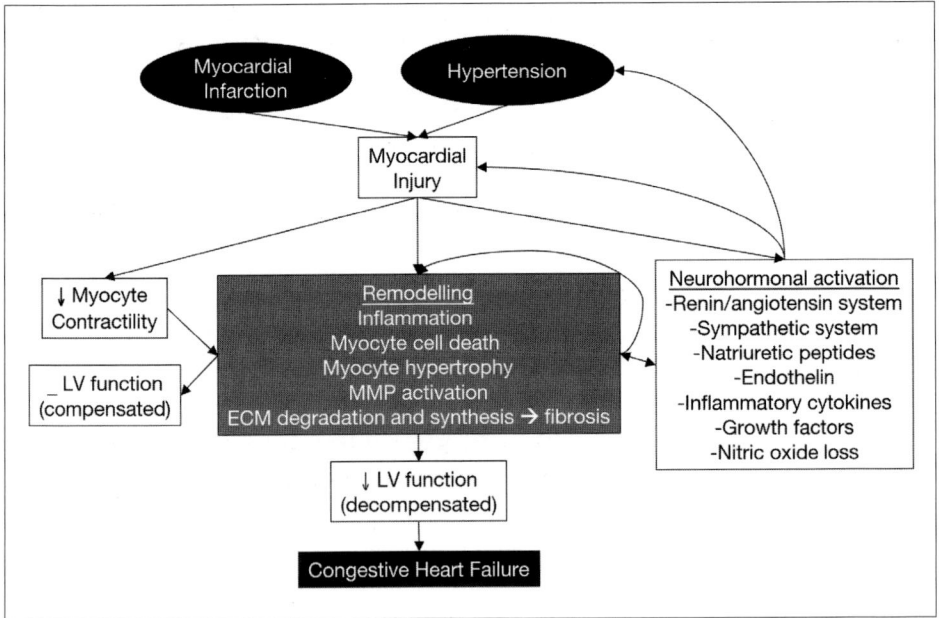

Figure 1

Both myocardial infarction and hypertension induce cardiac remodeling, which involves MMP activation, inflammation, cardiomyocyte cell death and hypertrophy, and extracellular matrix degradation and synthesis. Neurohormonal activation involves upregulation of the renin/angiotensin system, the sympathetic system, natriuretic peptides, endothelin, inflammatory cytokines, and growth factors, along with the loss of nitric oxide. While the response is initially compensatory, in time LV function decompensates and congestive heart failure is initiated.

remodeling as a part of the healing process. This phase involves an inflammatory response initiated by neutrophil infiltration. The initial response induces an increase in wall stress, which is then offset by an increase in LV mass through hypertrophy of still viable cardiomyocytes. Collagen synthesis is stimulated to induce fibrosis and MMP levels are increased to stimulate extracellular matrix turnover. LV mass is further increased, leading to an increase in chamber volume (measured by end systolic and diastolic dimensions), which ultimately impairs cardiac function. This begins the path of progressive remodeling that will, if not prevented, end in heart failure. Regardless of current therapeutic strategies, including reperfusion and anti-hypertensive agents, adverse cardiac remodeling progressing to CHF remains a leading cause of cardiovascular morbidity and mortality [10, 11]. A major challenge for current research efforts is to develop improved strategies to prevent, detect, or reverse adverse cardiac remodeling, because adverse remodeling becomes both a consequence and cause of CHF [8].

CHF is clinically defined as an abnormality in cardiac function that causes the heart to fail to pump blood at a rate necessary to adequately perfuse metabolizing tissues or when the heart can do so only at an elevated pressure [12]. In response to the index event (e.g., myocardial infarction or hypertension) the pumping capacity of the LV decreases, which in turn induces a secondary damage that further stimulates cardiac remodeling.

Matrix metalloproteinases (MMPs) are enzymes that regulate cardiac remodeling

The MMP family is composed of 25 individual members divided into specific classes based on *in vitro* substrate specificity for various extracellular matrix components. There are two major types of MMPs: secreted and membrane bound. Most MMPs represent the first group and are secreted as a proenzyme. For these MMPs, a primary regulation step is the activation of the proenzyme by cleavage of an approximately 10 kD propeptide in the amino terminus. Of the membrane type MMPs, MMP-14 has been most linked with cardiac remodeling [13]. MMP activity is inhibited non-specifically in the plasma by α_2 macroglobulin and specifically in the tissue by the tissue inhibitors of metalloproteinases (TIMPs), a family currently composed of four members [14]. TIMP-1 is expressed by many cells involved in cardiac remodeling and is positively association with disease progression [15]. TIMP-4 has high constitutive expression in the myocardium, but TIMP-4 is not cardiac specific in that many other cell types (e.g., platelets) also express TIMP-4 at high levels [16, 17]. *In vivo*, the most probable MMP activators are serine proteases and other MMPs, particularly MMP-3 [18] and the membrane type (MT) family members [19, 20]. While a clear case for MMP involvement in CHF has been made, Table 1 illustrates the fact that only 11 of the 25 MMPs have been evaluated in cardiovascular disease, myocardial infarction, hypertension, CHF, and/or cardiac remodeling. Whether the other 14 MMPs are differentially expressed during cardiovascular disease has not been determined. In contrast, all except four of the 25 MMPs have been identified in at least one cancer type.

MMP levels are induced by a variety of growth factors [21], cytokines [22, 23], and hormones [24]. In turn, MMPs process a wide range of ECM and non-ECM substrates. Non-ECM substrates include other proteinases, proteinase inhibitors, clotting factors, chemotactic molecules, latent growth factors, growth factor-binding proteins, cell surface receptors, and cell–cell adhesion molecules [25]. The list of MMP substrates may also include intracellular targets [26, 27]. MMP-23, for example, lacks a recognizable signal sequence, indicating that it is either intracel-

Table 1 - The association between MMPs and cardiovascular diseases. Only 11 of the 25 MMPs have been linked to cardiovascular disease.

MMP	CVD	MI	HTN	CHF	CR
1	yes	yes	yes	yes	yes
2	yes	yes	yes	yes	yes
3	yes	yes	yes	yes	yes
7	yes	yes	yes	yes	yes
8	yes	yes	yes	yes	yes
9	yes	yes	yes	yes	yes
10	yes	unk	unk	unk	unk
11	unk	unk	unk	unk	unk
12	yes	yes	yes	unk	yes
13	yes	yes	yes	yes	yes
14	yes	yes	unk	yes	yes
15	unk	unk	unk	unk	unk
16	unk	unk	unk	unk	unk
17	unk	unk	unk	unk	unk
18	unk	unk	unk	unk	unk
19	yes	unk	unk	unk	unk
20	unk	unk	unk	unk	unk
21	unk	unk	unk	unk	unk
22	unk	unk	unk	unk	unk
23	unk	unk	unk	unk	unk
24	unk	unk	unk	unk	unk
25	unk	unk	unk	unk	unk
26	unk	unk	unk	unk	unk
27	unk	unk	unk	unk	unk
28	unk	unk	unk	unk	unk
TOTAL	11	9	8	8	9

CVD, cardiovascular disease; MI, myocardial infarction; HTN, hypertension; CHF, congestive heart failure; CR, cardiac remodeling; yes, a link has been made; unk, unknown

lularly localized or that it utilizes an alternative secretory mechanism [28]. MMP-23 has not been evaluated in cardiac remodeling.

MMPs regulate multiple cell functions, including migration, invasion, proliferation, and apoptosis [25, 29]. MMPs regulate physiological processes such as angiogenesis and wound healing [30]. Most MMPs are expressed at low levels in

Table 2 - MMPs produced by the major cell types involved in cardiac remodeling

Cell type	MMPs
Cardiomyocyte	1, 2, 3, 7, 9, 14
Cardiac fibroblast	1, 2, 7, 9, 13, 14
Endothelial cells	1, 2, 3, 7, 8, 9, 10, 14
Vascular smooth muscle cells	1, 2, 3, 8, 9, 12, 13, 14
Neutrophils	8, 9, 25
Macrophages	1, 3, 7, 8, 9, 12, 13, 14

the myocardium and require a stimulus such as tissue injury or growth factor signaling [31]. Many MMPs are induced through activation of c-fos and c-jun, which heterodimerize and bind activator protein-1 on the MMP gene [25]. One exception to this rule is MMP-2, which is constitutively expressed at high levels in most tissues and appears to serve a housekeeping function.

The primary cells that contribute MMPs during cardiac remodeling are the resident cells (cardiac myocytes and fibroblasts), the coronary vasculature (endothelial and vascular smooth muscle cells), and inflammatory cells (neutrophils and macrophages). The MMPs known to be produced by each cell type are shown in Table 2. MMP-2 is constitutively expressed in the myocardium and likely plays a role in maintaining extracellular matrix homeostasis. Cardiomyocytes produce MMP-1, MMP-2, MMP-3, MMP-7, MMP-9, and MMP-14 [32, 33]. Cardiac fibroblasts produce MMP-1, MMP-2, MMP-7, MMP-9, MMP-13, and MMP-14 [33]. There is an interesting interplay between fibroblast secretory factors and MMPs. For examples, several MMPs (including MMP-1, 3, 7, 9, 13 and 14) have been shown to proteolytically process several fibroblast-derived factors, including transforming growth factor (TGF)-β and connective tissue growth factor (CTGF) [34–37]. TGF-β and CTGF in turn are also able to stimulate synthesis of MMP-1, MMP-2, MMP-3, MMP-9, MMP-12 and MMP-14, to decrease levels of TIMP-1 and TIMP-2, and to affect fibroblast chemoattraction and adhesion [38–41]. CTGF is also induced by tumor necrosis factor (TNF)-α, which increases MMPs and decreases TIMPs in fibroblasts and/or vascular endothelial cells [38–40].

Endothelial cells produce MMP-1, MMP-2, MMP-3, MMP-7, MMP-8, MMP 9, MMP-10 and MMP-14, while vascular smooth muscle cells produce MMP-1, MMP-2, MMP-3, MMP-8, MMP-9, MMP-12, MMP-13 and MMP-14. The fact that so many different MMP types are expressed by one cell suggests that, while some functions may be overlapping, distinct functions necessitate the expression of distinct MMPs. Neutrophils express MMP-8, MMP-9 and MMP-25, while macrophages express MMP-7, MMP-8, MMP-9, MMP-12, MMP-13 and MMP14 [33]. Leukocyte infiltration is a necessary component of wound healing that can also exacerbate the healing process. Different MMP roles in wound healing may explain the beneficial and negative aspects of leukocyte infiltration. When fibroblasts are activated in conjunction with macrophages, the fibroblast secretes MMPs, growth factors, and cytokines. Macrophages coordinate the differentiation of fibroblasts into myofibroblast and stimulate blood vessel formation by releasing angiogenic factors [42]. Together, these six cell types work in concert to regulate the cardiac response to injury, and MMPs are integral members of this response.

Remodeling induced by myocardial infarction

Myocardial infarction (MI) occurs when a previously atherosclerotic plaque ruptures, resulting in the downstream occlusion of the affected coronary artery. Post-MI wound healing involves the integration of biochemical, immunological, and structural changes that culminate in the *de novo* synthesis of scar tissue [43–46]. Mechanisms that maintain cardiac output post-MI also increase end diastolic volume, end diastolic pressure, and wall stress, which lead to dilation in the infarct region and compensatory hypertrophy in the remote region. The infarct scar is dynamic, comprised of cellular components that make up a metabolically active and vascularized tissue [47]. During the healing phase, infarct tissue is repaired and remodeled through extracellular matrix degradation and synthesis, angiogenesis, apoptosis, and proliferation that combine in a temporally regulated manner [37]. Infarct wall thinning amplifies the increase in wall stress, which further impairs LV function [45].

It was shown in the 1970s before reperfusion, repeated in the late 1980s with reperfusion, and shown again just a few years ago with updated therapies that the extent of remodeling post-MI, as indicated by dilation, is an independent determinant of survival [48–50]. Regardless of improved reperfusion strategies and the use of therapeutic strategies such as ACE inhibitors and β adrenergic receptor blockade, cardiac remodeling continues to impair a significant number of patients [51, 52]. An estimated 6–40% of post-MI patients will progress to CHF [53–58]. While the number of patients who progress to CHF has declined over the past 25 years, survival in patients diagnosed with CHF has not changed and the 5 year mortality rate remains at 50% [45, 59]. Regardless of variable incidence rates among studies, MI predisposes patients to CHF [60]. Identifying CHF early in the pathogenesis, therefore, will allow better risk stratification post-MI [60]. The traditional idea was that excessive MMP activation lead to increased extracellular matrix degradation, which in turn caused myocytes to slip and the LV wall to dilate and thin [12]. We now know that MMP functions in the post-MI and CHF myocardium are much more complex and involve both negative and positive roles.

While MMP-2 and MMP-9 are historically the easiest MMPs to study, due to their ability to be resolved on gelatin zymograms, multiple MMPs increase in response to MI and changes in several MMPs correlate with LV functional changes. Now that antibodies for most if not all MMPs are commercially available, future research examining all MMPs during cardiac remodeling should fill in the knowledge gaps that currently exist. MMPs have been evaluated in mouse, rat, sheep, dog, and rabbit models of MI, as well as in human patients [7, 61]. Karl Weber's group was the first to demonstrate that MMP levels increased following MI [62]. In human and non-rodent MI models, MMP-1 levels increase during the acute phase and decrease during the chronic phase [63]. The other major collagenases, MMP-8, MMP-13, and MMP-14, are all increased 8 weeks post-MI in rat and sheep models [63, 64]. MMP-2 and 3 levels increase slowly over time [32, 33, 63–66]. In rats, MMP-2 levels in the remote region differentiated between compensated and decompensated function at 3 months post-MI [33]. MMP-7 levels also increase post-MI [33, 63], and this increase occurs in the infarct region due to increased macrophage numbers and in the remote region due to upregulation in cardiomyocytes [67]. MMP-9 expression has been documented during all phases of post-MI remodeling [33, 64]. MMP-9 levels increase following myocardial ischemia and reperfusion, predominantly due to the influx of neutrophils [68]. MMP-9 levels also rise post-MI in the absence of reperfusion [32, 63, 65, 66], and MMP-9 plasma levels predict post-MI progression to CHF in patients [69]. By multivariate analysis, MMP-9 was the only early circulating factor evaluated that predicted late onset CHF, with an odds ratio of 6.5 ($p < 0.01$). TIMP-1 levels increase early and then fall off if the healing is compensated [33], whereas TIMP-2, TIMP-3, and TIMP-4 levels remain low at multiple time points [32, 63, 64].

The observation that MMPs are overexpressed during cardiac remodeling suggests a correlation but does not directly demonstrate a cause and effect relationship for these proteolytic enzymes in the development of LV dilation. MMP inhibition post-MI has been examined in both animal models and one clinical trial [70]. MMP inhibition using a broad spectrum MMP inhibitor (MMPi) attenuates LV dilation in mouse, rat, pig, and rabbit models of MI [71–76]. Infarcted mice were randomized to a broad-spectrum MMPi 24 h post-MI had significantly smaller increases in end-systolic and end-diastolic dimensions and areas at both mid-papillary and apical levels at the 4 day follow-up, compared with mice given the placebo [76]. In an analysis stratified by baseline end-diastolic area, the effects of MMP inhibition on the change in end-systolic area and end-diastolic area were most prominent in animals that had more initial left ventricular dilatation. Similar results were seen in a rabbit MI model, using an MMPi that did not inhibit MMP-1 [71]. Both MMP-2 and MMP-9 null mice are protected from LV rupture post-MI, suggesting a role for gelatinolytic activity in regulating adequate scar formation [77]. Survival post-MI was improved in MMP-7 null mice, due to improved electrical remodeling [67]. The absence of TIMP-1 exacerbates cardiac remodeling post-MI in mice, as evidenced by a greater increase in end diastolic volume [78]. In humans, MMP inhibition started at 48 h post-MI did not yield additional benefit over current optimal therapeutic options, indicating the need for further understanding of the basic MMP roles in cardiac remodeling before this MMP inhibition could be successfully adopted [79].

Remodeling induced by hypertension

Hypertension is defined as the chronic elevation of systolic blood pressure over 140 mmHg and diastolic blood pressure over 90 mmHg. Hypertension induces pressure overload on the myocardium by increasing aortic resistance to flow (after-load), which forces the LV to contract harder in order to eject the same volume of blood. In response to pressure overload, the initial response of the myocardium is hypertrophic, with cardiac myocyte growth occurring in a concentric manner such that the LV wall thickness to chamber volume ratio is increased. Concentric LV hypertrophy reduces wall stress and preserves LV function in the short-term, which allows ejection fraction and stroke volume to be maintained. Prolonged pressure overload can induce further structural changes, which can alter diastolic function and in time lead to diastolic heart failure. Chronic hypertension leads to an excess accumulation of extracellular matrix, which both leads to and is a consequence of myocyte loss. Further hypertrophy of the still viable myocytes occurs as a compensation mechanism, until maximum myocyte size limitations are reached [80]. When myocyte hypertrophy exceeds coronary vascular growth, tissue necrosis occurs [81]. In cardiac myocytes, the hypertrophic pathways involves several signaling molecules, including AKT and ERK 1/2 [80] NFAT and GATA-4 together activate the transcription of genes involved in the hypertrophic response [80]. In the case of chronic pressure overload that occurs over decades, the myocardium is additionally subjected to changes that normally occur as a result of the aging process [82, 83].

Multiple MMPs increase in response to pressure overload. Studies evaluating serum levels are confounded by the fact that clotting induces the release of MMPs from circulating leukocytes, which artificially elevates MMP measurements [84, 85]. Studies evaluating serum MMP levels, therefore, are not included here. MMP 2 and MMP-9 are also the most widely studied MMPs in hypertension studies, and both are elevated during the transition from compensatory hypertrophy to heart failure [86, 87]. In the Dahl high salt acute model of hypertension-induced hypertrophy and CHF, MMP-2, TIMP-2, and TIMP-4 levels increased during the transition from hypertrophy to failure [88]. MMP-2 and MMP-9 increased before LV dilation, and LV dilation stimulated further induction of these two MMPs [89]. In the spontaneously hypertensive rat model, MMP-13, TIMP-1, and TIMP-2 levels were also increased, and TIMP-4 levels were decreased [87]. In this study, two groups of rats were treated with either an angiotensin converting enzyme inhibitor (quinapril) or a MMP inhibitor (PD166793) for 4 months, beginning at 9 months of age when the transition from hypertrophy to heart failure begins, resulted in decreased LV dilation, preserved systolic function, and normalized MMP and TIMP levels. Quinapril, but not PD166793, decreased collagen mRNA levels and myocyte hypertrophy, indicating that different mechanisms of action explained the effects on MMP levels and LV function. In humans, MMP-1 levels decrease and TIMP-1 levels increase in the plasma of hypertensive patients compared to normotensive controls [90]. Treatment with the angiotensin converting enzyme inhibitor lisinopril for one year increased MMP-1 and lowered TIMP-1 levels. MMP-9 null mice show attenuated LV dysfunction following pressure overload [91], whereas TIMP-3 null mice show an exacerbated hypertrophic response following aortic banding model of pressure overload [77]. MMP-7 has also been shown to be involved in regulating G-protein coupled receptor-mediated hypertension in response to adrenergic receptor stimulation. Shedding of the heparin-binding epidermal growth factor from vascular cells has been linked to MMP-7 activity [89]. The MMP inhibitors GM6001 and doxycycline both prevented adrenergic-stimulated vasoconstriction. In the spontaneous hypertensive rat (but not the normotensive control), doxycycline treatment lowered systolic blood pressure and prevented heparin-binding epidermal growth factor shedding in mesenteric arteries [89]. The MMP inhibitor KB-R7785 prevented angiotensin-II induced epidermal growth factor receptor transactivation and prevented the hypertrophic response [89]. While MMP inhibition strategies have not been as well developed for hypertension as for post-MI remodeling, research in this area supports the concept of developing specific and selective MMP inhibitors for the treatment during particular phases of hypertensive remodeling. Because the MMP/TIMP balance is tipped in favor of extracellular matrix accumulation during the early response to hypertension, MMP activation and/or TIMP inhibition may be a better strategy during the acute hypertension phase. This concept remains to be fully developed.

MMP roles in adverse remodeling that progresses to CHF

MMP levels are generally increased in CHF [92]. MMP-1, MMP-2, MMP-3 and MMP-9 have each been documented to increase in CHF patients [93]. MMP-9 levels were also elevated in the plasma of CHF patients [94]. Decreased plasma TIMP-3 levels paralleled adverse remodeling in both a cardiomyopathic hamster model as well as human CHF [95]. Cardiac fibrosis, the net accumulation of extracellular matrix components in the LV, is a major contributor of CHF. Increased fibrosis leads to increased LV stiffness, which impairs normal function and stimulates further remodeling. While the cardiac extracellular matrix is composed of structural proteins (collagen and elastin), adhesive proteins (laminin, fibronectin, collagen IV), anti-adhesive proteins (osteopontin, tenascin, and thrombospondin), and proteoglycans (heparan sulfate) [96], collagen I is the most abundant and frequently studied extracellular matrix protein. Fibrosis is classified into two groups: reparative and reactive. Reparative fibrosis is the replacement of non-viable myocytes with extracellular matrix to form a scar. Reactive fibrosis occurs within the adventitia of intra-myocardial coronary arteries and arterioles (perivascular) or within the intermyo-cyte spaces (interstitial) [37]. Fibrosis is determined primarily by the proliferative and secretory capacity of the cardiac fibroblast [97–99]. Angiotensin II, endothelin, and aldosterone are each stimulated by pressure overload, each stimulate increased MMP production [100], and each stimulate excess collagen deposition in the LV [96]. Increasing levels of collagen, fibronectin, and TGF-β parallel the development of heart failure in the spontaneously hypertensive rat model and human CHF [96]. While TGF-β is generally considered to be a pro-fibrotic cytokine in the myocar-dium, cardiac fibroblasts respond to TGF-β stimulation by increasing MMP produc-tion and cell migration rates [101, 102].

Both interleukin (IL)-1β and TNF-α induce angiotensin II receptor I transcrip-tion in cardiac fibroblasts, while angiotensin II, TGF-β_1, and basic fibroblast growth factor all decreased receptor levels and norepinephrine, endothelin-1, atrial natriuretic peptide, and bradykinin had no significant effect [103]. TGF-β and fibroblast growth factor both stimulate fibroblasts to express other extracellular matrix proteins, including osteopontin and osteonectin, both of which influence fibrosis [104–106]. Osteopontin regulates fibroblast adhesion, growth, migration, and extracellular matrix synthesis. Osteonectin regulates collagen deposition [107]. Osteonectin cleavage by MMP-2, MMP-3, MMP-7, MMP-9, or MMP-13 increases the affinity of osteonectin for collagen I, collagen IV, and collagen V 7–20 fold [108]. Osteonectin cleavage by MMP-3 also generates peptides that regulate angio-genesis by increasing endothelial cell proliferation and migration [109].

Cardiac remodeling is an adaptive consequence of both the initial injury insult and the ensuing compensation-induced damage. A spectrum of mechanisms com-prise the signs and symptoms of CHF, including neurohormonal activation, altered cardiac myocyte function, changes in LV chamber geometry, and cardiac remodeling [12]. In addition to increased MMP levels, CHF progression has also been linked to alterations in myocardial stretch, excess norepinephrine stimulation, TNF-α, oxida-tive stress, and angiotensin II [12]. Compensatory mechanisms include adrenergic and sympathetic nervous system activation, salt and water retaining systems (e.g., renin-angiotensin-aldosterone system), activation of vasodilators (e.g., natriuretic peptides, prostaglandins such as PGE2 and PGEI2, and nitric oxide), and induction of the cytokine system [12]. Norepinephrine, angiotensin II, endothelin-1, aldoste-rone, and TNF-α are biologically active molecules, each of which could contribute to CHF progression [110–115]. Norepinephrine, angiotensin II, endothelin, and tumor necrosis factor are synthesized as autocrine and paracrine factors directly in the myocardium [12]. The neurohormonal model has been proposed to explain why CHF can develop years after the precipitating event and why the phenotype is consistently uniform despite etiological differences [12].

Circulating vasoactive peptides contribute to cardiovascular hemodynamic function by regulating vascular tone, sodium retention, and intravascular volume

[116]. Sodium regulating hormones include catecholamines, endothelin-1, natriuretic peptides, nitric oxide, angiotensin II, and aldosterone. These hormones are each systemically elevated in proportion to the degree of heart failure. Involvement of the renin-angiotensin-aldosterone system in all stages of MI, hypertension, and cardiac remodeling are well documented [117–120], and this involvement includes the upregulation of MMPs [116]. Natriuretic peptides mediate the early compensatory response, stimulating sodium and water retention and vasoconstriction. Brain natriuretic peptide is produced by cardiac fibroblasts, and increased levels induce MMP-1, MMP-2, MMP-3, and MMP-14 in cardiac fibroblasts through cGMP protein kinase G signaling [121]. Endothelin is a potent vasoconstrictor that is induced by angiotensin II, norepinephrine, or mechanical stretch, indicating an interplay between these bioactive molecules [89]. Catecholamines, angiotensin II, and endothelin-1 each promote fluid retention in the kidneys, increase vasoconstriction, and stimulate cardiac contractility. Vasoconstrictors, however, usually also promote fibrosis. Norepinephrine, endothelin-1, and angiotensin II stimulate fibroblast proliferation and collagen synthesis directly through the upregulation of c-fos and c-jun which heterodimerize and bind to activator protein-1 [122]. Vasodilators, in contrast, tend to be anti-fibrotic and inhibit both fibroblast proliferation and collagen synthesis. Natriuretic peptides and nitric oxide work by activating cGMP that then activates protein kinase G [123].

While these peptides attempt to restore physiologic homeostasis, each can further exacerbate heart failure through direct action on cardiomyocyte and fibroblast function. For example, cardiac fibroblasts both secrete and respond to brain natriuretic peptide (BNP) [121]. TNF-α stimulated BNP release, and the addition of BNP to cardiac fibroblast cultures inhibited collagen synthesis and increased protein expression of MMPs 1, 2, 3 and 14. Increased BNP, therefore, would stimulate collagen turnover and LV dilation.

Angiotensin II can also participate in several steps of inflammatory responses, including mediating oxidative stress, stimulating the release of cytokines and growth factors, and regulating the recruitment of pro-inflammatory cells into the site of injury [124]. Angiotensin II and aldosterone induce cardiac fibroblast proliferation and fibrosis through distinct mechanisms [125]. Angiotensin II stimulates the synthesis of both collagen and TGF-β_1 in adult cardiac fibroblasts and also can partially differentiate fibroblasts into myofibroblasts (increased expression of integrins and actin and increased myofilament organization [124]). While most studies looking at angiotensin II effects on cardiac fibroblast proliferation use neonatal fibroblasts, there are reports that angiotensin II can induce proliferation in adult cells [124].

Oxidative stress can induce changes associated with cardiac remodeling, including myocyte hypertrophy, apoptosis, fetal gene re-expression, and increased MMP activity [126]. Failing hearts produce more reactive oxygen species (ROS) and also show diminished antioxidant capacity such that the net effect is increased oxidative stress [126]. Mitochondria are the source of ROS due to electron leakage from the respiratory chain to form reactive oxygen [126]. In the myocardium, anti-oxidant enzymes protect cells by controlling superoxide anion and hydrogen peroxide levels. There are three types of superoxide dismutase (SOD), the enzyme that regulates ROS levels in the myocardium: manganese SOD, copper/zinc SOD, and extracellular SOD. The manganese SOD contributes approximately 90% of the SOD activity in cardiac myocytes and is located in the mitochondria. Mice deficient in the manganese SOD gene die soon after birth as a result of cardiomyopathy [127]. Copper/zinc SOD and extracellular SOD gene deficient mice show no overt cardiac phenotype in the absence of stress [128]. Glutathione peroxidase and catalase scavenge the hydrogen peroxide produced by the SOD enzymes. Both SOD and glutathione peroxidase activities decrease in heart failure, although these findings are somewhat controversial [126].

Increased oxidative stress induced by hydrogen peroxide decreases collagen synthesis and increases MMP-2, MMP-9, and MMP-13 levels in cardiac fibroblasts, both of which would favor dilation [129]. Several studies have demonstrated that reactive oxygen species induce cardiomyocyte death and that treatment of a post-MI mouse with the antioxidant DMTU reduces the extent of LV dilation by reducing MMP activity [126]. Reactive oxygen species work by activation the MAP Kinase pathway, including stimulating ERK and JNK [126]. Activation of p38α MAPK in turn stabilizes MMP-1 and MMP-3 mRNA, leading to increased collagenolytic potential [130].

Current treatment modalities for CHF and their effects on MMP levels

Current CHF management strategies primarily involve relieving symptoms, slowing progression, and reversing structural abnormalities [131]. The main symptoms are dyspnea (shortness of breath), decreased exercise tolerance, lower extremity edema (fluid accumulation), and organ dysfunction (including renal failure). The major symptomatic endpoint to measure treatment efficacy is improved exercise tolerance, and the major survival endpoint to measure treatment efficacy is increased ejection fraction. Despite biological heterogeneity in response, current hypertensive, post-MI, and CHF inhibition therapies are, for the most part, universal, with all patients receiving the same treatment [132].

Current pharmacological agents that treat symptoms include diuretics, vasodilators (ACE inhibitors), positive inotropes, and digoxin. Agents that treat LV dysfunction inhibit the neurohormonal system and remodeling and include angiotensin converting enzyme inhibitors, nitrates, hydralazine, and β adrenergic receptor inhibitors. Blocking the aldosterone, angiotensin, endothelin, and norepinephrine each appear to positively influence remodeling [118]. ACE inhibitors decrease preload and afterload, but the effects of ACE inhibition are not entirely hemodynamic. Inhibiting the angiotensin converting enzyme improves cardiac function and decreases CHF mortality in both patients and animal models by reducing hypertension, ventricular hypertrophy, and fibrosis [119, 133, 134]. Doses of Ramipril that are too low to alter blood pressure reduce fibrosis, indicating that signaling mechanisms within the cardiac fibroblast may be responding to ACE inhibition [96]. Activation of the prostanoid, bradykinin, and nitric oxide signaling pathways have been proposed, because each can inhibit fibroblast proliferation and extracellular matrix synthesis *in vitro* [96]. ACE inhibitors also directly regulate MMP activation to inhibit excess extracellular matrix degradation [135].

Novel pharmacologic options

As has been shown with ACE inhibitors, drugs that alter extracellular matrix remodeling are likely to prove beneficial in the CHF patient. These include agents that alter collagen deposition and crosslinking, such as blockers of MMP function and advanced glycation end product inhibitors. This area will be difficult to develop, because extracellular matrix biology generally and MMP functions specifically are very complex, with multiple pathways regulating distinct and overlapping functions [25]. Because ACE inhibitors already exert a positive effect on cardiac remodeling, inhibitors that only incrementally add to this benefit will be hard to assess. Strategies that prevent extracellular matrix degradation and/or prevent excessive replacement fibrosis will also need to ensure that physiological wound healing is not also inhibited. The main issues that need to be worked out in order for future therapies to work are which MMP functions to inhibit and when is the optimal timing for inhibition. MMP inhibitor strategies have not produced straightforward results in both cancer and cardiovascular clinical trials, in part because we still do not have a firm understanding of the basic biological mechanisms that regulate remodeling [136, 137]. The balance between extracellular matrix synthesis and degradation needs to be determined such that the optimal ratio is achieved [138]. Excess extracellular matrix accumulation can lead to increased LV stiffness and dilation, while excess extracellular matrix degradation can lead to wall thinning and rupture. Understanding the fine controls that shift remodeling within the range between these extremes will provide mechanistic insight that can be used to more specifically target cardiac remodeling.

Future directions and conclusions

Challenges for the future include: 1) developing assays to predict who will develop CHF, especially methods to detect CHF at an earlier stage; 2) increasing interventions that will prevent the development or slow the progression of CHF; and 3) using biomarkers to guide treatment decisions (conservative *versus* aggressive treat-

Table 3 - Research directions to develop useful MMP inhibitor strategies

1. Establish a complete catalog of MMP changes during all stages of cardiac remodeling, from the precipitating event to advanced CHF.
2. Demonstrate cause and effect relationships in cardiac remodeling, for each MMP.
3. Delineate MMP proteolytic events that are unique to adverse cardiac remodeling.
4. Determine what MMP-induced changes are necessary for wound healing responses and which changes are potentially negative side effects (both intra-cardiac and systemic).
5. Develop selective inhibitors that only block specific MMPs and/or specific MMP functions.

ment strategies) [132]. Each of these challenges involves further research on MMP roles. MMP research in the cardiac system, especially in terms of MMP roles during cardiac remodeling leading to heart failure, has not been saturated (Tab. 3). A complete catalog of which MMPs change, where the change is localized, and why the change occurs is still needed. Additionally, a complete catalog of substrates proteolyzed by MMPs has not been developed. MMPs have both positive and negative roles in cardiac remodeling, and isolating out the negative without impacting the positive will provide novel therapeutic tools for cardiac remodeling.

Acknowledgements

The authors acknowledge the following grant support from the National Institutes of Health: HL-75360 (MLL).

References

1 Cohn JN, Ferrari R, Sharpe N (2000) Cardiac remodeling – concepts and clinical implications: a consensus paper from an International Forum on Cardiac Remodeling. *J Am Coll Cardiol* 35: 569–582

2 Pfeffer JM, Pfeffer MA, Fletcher PJ, Braunwald F. (1991) Progressive ventricular remodeling in rat with myocardial infarction. *Am J Physiol* 260: H1406–H1414

3 Pfeffer MA, Braunwald E (1990) Ventricular remodeling after myocardial infarction. Experimental observations and clinical implications. *Circulation* 81: 1161–1172

4 Cohn JN, Ferrari R, Sharpe N (2000) Cardiac remodeling – concepts and clinical implications: a consensus paper from an international forum on cardiac remodeling. *J Am College Cardiol* 35: 569–582

5   Ren G (2003) Inflammatory mechanisms in myocardial infarction. *Curr Drug Targets – Inflammation & Allergy* 2: 242–256

6   Frangogiannis NG, Smith CW, Entman ML (2002) The inflammatory response in myocardial infarction. *Cardiovasc Res* 53: 31–47

7   Lindsey ML (2004) MMP induction and inhibition in myocardial infarction. *Heart Fail Rev* 9: 7–19

8   Kurrelmeyer K, Kalra D, Bozkurt B, Wang F, Dibbs Z, Seta Y, Baumgarten G, Engle D, Sivasubramanian N, Mann DL (1998) Cardiac remodeling as a consequence and cause of progressive heart failure. *Clin Cardiol* 21: I14–I19

9   Janicki JS, Brower GL, Gardner JD, Chancey AL, Stewart Jr JA (2004) The dynamic interaction between matrix metalloproteinase activity and adverse myocardial remodeling. *Heart Fail Rev* 9: 33–42

10 Yan AT, Yan RT, Liu PP (2005) Narrative review: pharmacotherapy for chronic heart failure: evidence from recent clinical trials. *Ann Intern Med* 142: 132–145

11 Azevedo CF, Cheng S, Lima JA (2005) Cardiac imaging to identify patients at risk for developing heart failure after myocardial infarction. *Curr Heart Fail Rep* 2: 183–188

12 Mann DL (1999) Mechanisms and models in heart failure: a combinatorial approach. *Circulation* 100: 999–1008

13 Manso AM, Elsherif L, Kang S-M, Ross RS (2006) Integrins, membrane-type matrix metalloproteinases and ADAMs: Potential implications for cardiac remodeling. *Cardiovasc Res* 69: 574–584

14 Brew K, Dinakarpandian D, Nagase H (2000) Tissue inhibitors of metalloproteinases: evolution, structure and function. *Biochim Biophys Acta* 1477: 267–283

15 Cavusoglu E, Ruwende C, Chopra V, Yanamadala S, Eng C, Clark LT, Pinsky DJ, Marmur JD (2006) Tissue inhibitor of metalloproteinase-1 (TIMP-1) is an independent predictor of all-cause mortality, cardiac mortality, and myocardial infarction. *Am Heart J* 151: 1101.e1101–1101.e1108

16 Leco KJ, Apte SS, Taniguchi GT, Hawkes SP, Khokha R, Schultz GA, Edwards DR (1997) Murine tissue inhibitor of metalloproteinase-4 (Timp-4): cDNA isolation and expression in adult mouse tissues. *FEBS Letters* 401: 213–217

17 Liu YE, Wang M, Greene J, Su J, Ullrich S, Li H, Sheng S, Alwxander P, Sang QA, Shi YE (1997) Preparation and characterization of recombinant tissue inhibitor of metalloproteinase 4 (TIMP-4). *J Biol Chem* 272: 20479–20483

18 Murphy G, Cockett MI, Stephens PE, Smith BJ, Docherty AJP (1987) Stromelysin is an activator of procollagenase. *Biochem J* 248: 265–268

19 Cao J, Drews M, Lee HM, Conner C, Bahou WF, Zucker S (1998) The propeptide domain of membrane type 1 matrix metalloproteinase is required for binding of tissue inhibitor of metalloproteinases and for activation of pro-gelatinase A. *J Biol Chem* 273: 34745–34752

20 Cowell S, Knquper V, Stewart ML, D'Ortho M-P, Stanton H, Hembry RM, Lopez-Otin C, Reynolds JJ, Murphy G (1998) Induction of matrix metalloproteinase activation

cascades based on membrane-type 1 matrix metalloproteinase: associated activation of gelatinase A, gelatinase B and collagenase 3. *Biochem J* 331: 453–458

21   Janowska-Wieczorek A, Marquez LA, Nabholtz JM, Cabuhat ML, Montano J, Chang H, Rozmus J, Russell JA, Edwards DR, Turner AR (1999) Growth factors and cytokines upregulate gelatinase expression in bone marrow CD34(+) cells and their transmigration through reconstituted basement membrane. *Blood* 93: 3379–3390

22   Ries C, Petrides PE (1995) Cytokine regulation of matrix metalloproteinase activity and its regulatory dysfunction in disease. *Biol Chem Hoppe Seyler* 376: 345–355

23   Mauviel A (1993) Cytokine regulation of metalloproteinase gene expression. *J Cell Biochem* 53: 288–295

24   Benbow U, Brinckerhoff CE (1997) The AP-1 Site and MMP gene regulation: what is all the fuss about? *Matrix Biol* 15: 519–526

25   Sternlicht M, Werb Z (2001) How matrix metalloproteinases regulate cell behavior *Annu Rev Cell Dev Biol* 17: 463–516

26   Van den Steen PE, Van Aelst I, Hvidberg V, Piccard H, Fiten P, Jacobsen C, Moestrup SK, Fry S, Royle L, Wormald MR et al (2006) The hemopexin and O-glycosylated domains tune gelatinase B/MMP-9 bioavailability *via* inhibition and binding to cargo receptors. *J Biol Chem* 281: 18626–18637

27   Sawicki G, Leon H, Sawicka J, Sariahmetoglu M, Schulze CJ, Scott PG, Szczesna-Cordary D, Schulz R (2005) Degradation of myosin light chain in isolated rat hearts subjected to ischemia-reperfusion injury: a new intracellular target for matrix metallo-proteinase-2. *Circulation* 112: 544–552

28   Velasco G, Pendas AM, Fueyo A, Knauper V, Murphy G, Lopez-Otin C (1999) Cloning and characterization of human MMP-23, a new matrix metalloproteinase predomi-nantly expressed in reproductive tissues and lacking conserved domains in other family members. *J Biol Chem* 274: 4570–4576

29   Egeblad M, Werb Z (2002) New functions for the matrix metalloproteinases in cancer progression. *Nat Rev Cancer* 2: 163–176

30   Vu TH, Werb Z (2000) Matrix metalloproteinases: effectors of development and normal physiology. *Genes Dev* 14: 2123–2133

31   Nelson AR, Fingleton B, Rothenberg ML, Matrisian LM (2000) Matrix metalloprotein-ases: biologic activity and clinical implications. *J Clin Oncol* 18: 1135–1149

32   Romanic AM, Burns-Kurtis CL, Gout B, Berrebi-Bertrand I, Ohlstein EH (2001) Matrix metalloproteinase expression in cardiac myocytes following myocardial infarction in the rabbit. *Life Sci* 68: 799–814

33   Gaertner R, Jacob M-P, Prunier F, Angles-Cano E, Mercadier J-J, Michel J-B (2005) The plasminogen-MMP system is more activated in the scar than in viable myocardium 3 months post-MI in the rat. *J Mol Cell Cardiol* 38: 193–204

34   Overall CM, Tam EM, Kappelhoff R, Connor A, Ewart T, Morrison CJ, Puente X, Lopez-Otin C, Seth A (2004) Protease degradomics: mass spectrometry discovery of protease substrates and the CLIP-CHIP, a dedicated DNA microarray of all human proteases and inhibitors. *Biol Chem* 385: 493–504

35 Hashimoto G, Inoki I, Fujii Y, Aoki T, Ikeda E, Okada Y (2002) Matrix metalloproteinases cleave connective tissue growth factor and reactivate angiogenic activity of vascular endothelial growth factor 165. *J Biol Chem* 277: 36288–36295

36 Yu Q, Stamenkovic I (2000) Cell surface-localized matrix metalloproteinase-9 proteolytically activates TGF-beta and promotes tumor invasion and angiogenesis. *Genes Dev* 14: 163–176

37 Manabe I, Shindo T, Nagai R (2002) Gene expression in fibroblasts and fibrosis: involvement in cardiac hypertrophy. *Circ Res* 91: 1103–1113

38 Ivkovic S, Yoon BS, Popoff SN, Safadi FF, Libuda DE, Stephenson RC, Daluiski A, Lyons KM (2003) Connective tissue growth factor coordinates chondrogenesis and angiogenesis during skeletal development. *Development* 130: 2779–2791

39 Kondo S, Kubota S, Shimo T, Nishida T, Yosimichi G, Eguchi T, Sugahara T, Takigawa M (2002) Connective tissue growth factor increased by hypoxia may initiate angiogenesis in collaboration with matrix metalloproteinases. *Carcinogenesis* 23: 769–776

40 Chen C-C, Chen N, Lau LF (2001) The angiogenic factors Cyr61 and connective tissue growth factor induce adhesive signaling in primary human skin fibroblasts. *J Biol Chem* 276: 10443–10452

41 Shi-wen X, Stanton LA, Kennedy L, Pala D, Chen Y, Howat SL, Renzoni EA, Carter DE, Bou-Gharios G, Stratton RJ et al (2006) CCN2 is necessary for adhesive responses to transforming growth factor-beta1 in embryonic fibroblasts. *J Biol Chem* 281: 10715–10726

42 Chaponnier C, Gabbiani G (2004) Pathological situations characterized by altered actin isoform expression. *J Pathol* 204: 386–395

43 Rosado A, Lamas GA (1997) Left ventricular remodeling: clinical significance and therapy. *Basic Res Cardiol* 92: 66–68

44 Janicki JS, Brower GL, Henegar JR, Wang L (1995) Ventricular remodeling in heart failure: the role of myocardial collagen. *Adv Exp Med Biol* 382: 239–245

45 Yousef ZR, Redwood SR, Marber MS (2000) Postinfarction left ventricular remodeling: a pathophysiological and therapeutic review. *Cardiovasc Drugs Ther* 14: 243–252

46 Dietz R, Osterziel KJ, Willenbrock R, Gulba DC, von Harsdorf R (1999) Ventricular remodeling after acute myocardial infarction. *Thromb Haemost* 82 (Suppl 1): 73–75

47 Weber KT, Sun Y, Ratajska A, Cleutjens JPM, Tyagi SC (1995) Structural remodeling of the myocardium in ischemic and hypertensive heart disease. In: NS Dhalla, RE Beamish, N Takeda, M Nagano (eds): *The Failing Heart*. Lippincott-Raven Publishers, Philadelphia, 163–185

48 White HD, Norris RM, Brown MA, Brandt PW, Whitlock RM, Wild CJ (1987) Left ventricular end-systolic volume as the major determinant of survival after recovery from myocardial infarction. *Circulation* 76: 44–51

49 Gaudron P, Kugler I, Hu K, Bauer W, Eilles C, Ertl G (2001) Time course of cardiac structural, functional and electrical changes in asymptomatic patients after myocardial infarction: their inter-relation and prognostic impact. *J Am Coll Cardiol* 38: 33–40

50 Migrino RQ, Young JB, Ellis SG, White HD, Lundergan CF, Miller DP, Granger CB,

Ross AM, Califf RM, Topol EJ (1997) End-systolic volume index at 90 to 180 minutes into reperfusion therapy for acute myocardial infarction is a strong predictor of early and late mortality. The Global Utilization of Streptokinase and t-PA for Occluded Coronary Arteries (GUSTO)-I Angiographic Investigators. *Circulation* 96: 116–121

51 Solomon SD, Pfeffer MA (1997) The decreasing incidence of left ventricular remodeling following myocardial infarction. *Basic Res Cardiol* 92: 61–65

52 Bolognese L, Cerisano G (1999) Early predictors of left ventricular remodeling after acute myocardial infarction. *Am Heart J* 138: 79–83

53 Sutton MSJ, Pfeffer MA, Moye L, Plappert T, Rouleau JL, Lamas G, Rouleau J, Parker JO, Arnold MO, Sussex B et al (1997) Cardiovascular death and left ventricular remodeling two years after myocardial infarction: baseline predictors and impact of long-term use of captopril: information from the survival and ventricular enlargement (SAVE) trial. *Circulation* 96: 3294–3299

54 Solomon SD, Sutton MSJ, Lamas GA, Plappert T, Rouleau JL, Skali H, Moye L, Braunwald E, Pfeffer MA, for the Survival And Ventricular Enlargement (SAVE) Investigators (2002) Ventricular remodeling does not accompany the development of heart failure in diabetic patients after myocardial infarction. *Circulation* 106: 1251–1255

55 Lavine SJ (2003) Prediction of heart failure post myocardial infarction: comparison of ejection fraction, transmitral filling parameters, and the index of myocardial performance. *Echocardiography* 20: 691–701

56 Hellermann JP, Jacobsen SJ, Redfield MM, Reeder GS, Weston SA, Roger VL (2005) Heart failure after myocardial infarction: clinical presentation and survival. *Eur J Heart Fail* 7: 119–125

57 Hellermann JP, Goraya TY, Jacobsen SJ, Weston SA, Reeder GS, Gersh BJ, Redfield MM, Rodeheffer RJ, Yawn BP, Roger VL (2003) Incidence of heart failure after myocardial infarction: is it changing over time? *Am J Epidemiol* 157: 1101–1107

58 Lewis EF, Moye LA, Rouleau JL, Sacks FM, Arnold JM, Warnica JW, Flaker GC, Braunwald E, Pfeffer MA (2003) Predictors of late development of heart failure in stable survivors of myocardial infarction: the CARE study. *J Am Coll Cardiol* 42: 1446–1453

59 Levy WC, Mozaffarian D, Linker DT, Sutradhar SC, Anker SD, Cropp AB, Anand I, Maggioni A, Burton P, Sullivan MD et al (2006) The Seattle heart failure model: prediction of survival in heart failure. *Circulation* 113: 1424–1433

60 Anavekar NS, McMurray JJV, Velazquez EJ, Solomon SD, Kober L, Rouleau J-L, White HD, Nordlander R, Maggioni A, Dickstein K et al (2004) Relation between renal dysfunction and cardiovascular outcomes after myocardial infarction. *N Engl J Med* 351: 1285–1295

61 Vanhoutte D, Schellings M, Pinto Y, Heymans S (2006) Relevance of matrix metalloproteinases and their inhibitors after myocardial infarction: A temporal and spatial window. *Cardiovasc Res* 69: 604–613

62 Cleutjens JPM, Kandala JC, Guarda E, Guntaka RV, Weber KT (1995) Regulation of collagen degradation in the rat myocardium after infarction. *J Mol Cell Cardiol* 27: 1281–1292

63 Wilson EM, Moainie SL, Baskin JM, Lowry AS, Deschamps AM, Mukherjee R, Guy TS, St John-Sutton MG, Gorman JH, III, Edmunds LH Jr et al (2003) Region- and type-specific induction of matrix metalloproteinases in post-myocardial infarction remodeling. *Circulation* 107: 2857–2863

64 Peterson JT, Li H, Dillon L, Bryant JW (2000) Evolution of matrix metalloprotease and tissue inhibitor expression during heart failure progression in the infarcted rat. *Cardiovasc Res* 46: 307–315

65 Herzog E, Gu A, Kohmoto T, Burkhoff D, Hochman JS (1998) Early activation of metalloproteinases after experimental myocardial infarction occurs in infarct and noninfarct zones. *Cardiovasc Pathol* 7: 307–312

66 Tao Z-Y, Cavasin MA, Yang F, Liu Y-H, Yang X-P (2004) Temporal changes in matrix metalloproteinase expression and inflammatory response associated with cardiac rupture after myocardial infarction in mice. *Life Sciences* 74: 1561–1572

67 Lindsey ML, Escobar GP, Mukherjee R, Goshorn DK, Sheats NJ, Bruce JA, Mains IM, Hendrick JK, Hewett KW, Gourdie RG et al (2006) Matrix metalloproteinase-7 affects connexin-43 levels, electrical conduction, and survival after myocardial infarction. *Circulation* 113: 2919–2928

68 Lindsey M, Wedin K, Brown MD, Keller C, Evans AJ, Smolen J, Burns AR, Rossen RD, Michael L, Entman M (2001) Matrix-dependent mechanism of neutrophil-mediated release and activation of matrix metalloproteinase 9 in myocardial ischemia/reperfusion. *Circulation* 103: 2181–2187

69 Wagner DR, Delagardelle C, Ernens I, Rouy D, Vaillant M, Beissel J (2006) Matrix metalloproteinase-9 is a marker of heart failure after acute myocardial infarction. *J Cardiac Failure* 12: 66–72

70 Creemers EE, Cleutjens JP, Smits JF, Daemen MJ (2001) Matrix metalloproteinase inhibition after myocardial infarction: a new approach to prevent heart failure? *Circ Res* 89: 201–210

71 Lindsey ML, Gannon J, Aikawa M, Schoen FJ, Rabkin E, Lopresti-Morrow L, Crawford J, Black S, Libby P, Mitchell PG et al (2002) Selective matrix metalloproteinase inhibition reduces left ventricular remodeling but does not inhibit angiogenesis after myocardial infarction. *Circulation* 105: 753–758

72 Peterson JT, Hallak H, Johnson L, Li H, O'Brien PM, Sliskovic DR, Bocan TMA, Coker ML, Etoh T, Spinale FG (2001) Matrix metalloproteinase inhibition attenuates left ventricular remodeling and dysfunction in a rat model of progressive heart failure. *Circulation* 103: 2303–2309

73 Mukherjee R, Brinsa TA, Dowdy KB, Scott AA, Baskin JM, Deschamps AM, Lowry AS, Escobar GP, Lucas DG, Yarbrough WM et al (2003) Myocardial infarct expansion and matrix metalloproteinase inhibition. *Circulation* 107: 618–625

74 Yarbrough WM, Mukherjee R, Escobar GP, Mingoia JT, Sample JA, Hendrick JW, Dowdy KB, McLean JE, Lowry AS, O'Neill TP et al (2003) Selective targeting and timing of matrix metalloproteinase inhibition in post-myocardial infarction remodeling. *Circulation* 108: 1753–1759

75 Villarreal FJ, Griffin M, Omens J, Dillmann W, Nguyen J, Covell J (2003) Early short-term treatment with doxycycline modulates postinfarction left ventricular remodeling. *Circulation* 108: 1487–1492

76 Rohde LE, Ducharme A, Arroyo LH, Aikawa M, Sukhova GH, Lopez-Anaya A, McClure KF, Mitchell PG, Libby P, Lee RT (1999) Matrix metalloproteinase inhibition attenuates early left ventricular enlargement after experimental myocardial infarction in mice. *Circulation* 15: 3063–3070

77 Janssens S, Lijnen HR (2006) What has been learned about the cardiovascular effects of matrix metalloproteinases from mouse models? *Cardiovasc Res* 69: 585–594

78 Creemers EEJM, Davis JN, Parkhurst AM, Leenders P, Dowdy KB, Hapke E, Hauet AM, Escobar PG, Cleutjens JPM, Smits JFM et al (2003) Deficiency of TIMP-1 exacerbates LV remodeling after myocardial infarction in mice. *Am J Physiol Heart Circ Physiol* 284: H364–371

79 Hudson MP, Armstrong PW, Ruzyllo W, Brum J, Cusmano L, Krzeski P, Lyon R, Quinones M, Theroux P, Sydlowski D et al (2006) Effects of selective matrix metalloproteinase inhibitor (PG-116800) to prevent ventricular remodeling after myocardial infarction: results of the PREMIER (Prevention of Myocardial Infarction Early Remodeling) trial. *J Am College Cardiol* 48: 15–20

80 Wakatsuki T, Schlessinger J, Elson EL (2004) The biochemical response of the heart to hypertension and exercise. *Trends Biochem Sci* 29: 609–617

81 Tsotetsi OJ, Woodiwiss AJ, Netjhardt M, Qubu D, Brooksbank R, Norton GR (2001) Attenuation of cardiac failure, dilatation, damage, and detrimental interstitial remodeling without regression of hypertrophy in hypertensive rats. *Hypertension* 38: 846–851

82 Kass DA (2002) Age-related changes in venticular-arterial coupling: pathophysiologic implications. *Heart Fail Rev* 7: 51–62

83 Pepe S, Lakatta EG (2005) Aging hearts and vessels: Masters of adaptation and survival. *Cardiovasc Res* 66: 190–193

84 Gerlach RF, Demacq C, Jung K, Tanus-Santos JE (2007) Rapid separation of serum does not avoid artificially higher matrix metalloproteinase (MMP)-9 levels in serum *versus* plasma. *Clin Biochem* 40: 119–123

85 Thrailkill K, Cockrell G, Simpson P, Moreau C, Fowlkes J, Bunn RC (2006) Physiological matrix metalloproteinase (MMP) concentrations: comparison of serum and plasma specimens. *Clin Chem Lab Med* 44: 503–504

86 Yasmin, Wallace S, McEniery CM, Dakham Z, Pusalkar P, Maki-Petaja K, Ashby MJ, Cockcroft JR, Wilkinson IB (2005) Matrix metalloproteinase-9 (MMP-9), MMP-2, and serum elastase activity are associated with systolic hypertension and arterial stiffness. *Arterioscler Thromb Vasc Biol* 25: 372

87 Li H, Simon H, Bocan TMA, Peterson JT (2000) MMP/TIMP expression in spontaneously hypertensive heart failure rats: the effect of ACE- and MMP-inhibition. *Cardiovasc Res* 46: 298–306

88 Iwanaga Y, Aoyama T, Kihara Y, Onozawa Y, Yoneda T, Sasayama S (2002) Excessive activation of matrix metalloproteinases coincides with left ventricular remodeling dur-

ing transition from hypertrophy to heart failure in hypertensive rats. *J Am Coll Cardiol* 39: 1384–1391

89   Shah BH, Catt KJ (2004) Matrix metalloproteinase-dependent EGF receptor activation in hypertension and left ventricular hypertrophy. *Trends Endocrin Metab* 15: 241–243

90   Laviades C, Varo N, Fernandez J, Mayor G, Gil MJ, Monreal I, Diez J (1998) Abnormalities of the extracellular degradation of collagen type I in essential hypertension. *Circulation* 98: 535–540

91   Heymans S, Lupu F, Terclavers S, Vanwetswinkel B, Herbert J-M, Baker A, Collen D, Carmeliet P, Moons L (2005) Loss or inhibition of uPA or MMP-9 attenuates LV remodeling and dysfunction after acute pressure overload in mice. *Am J Pathol* 166: 15–25

92   Lee RT (2001) Matrix metalloproteinase inhibition and the prevention of heart failure. *Trends Cardiovasc Med* 11: 202–205

93   Coker ML, Thomas CV, Clair MJ, Hendrick JW, Kromback RS, Galis ZS, Spinale FG (1998) Myocardial matrix metalloproteinase activity and abundance with congestive heart failure. *Am J Physiol Heart Circ Physiol* 274: H1516–H1523

94   Wilson EM, Gunasinghe HR, Coker ML, Sprunger P, Lee-Jackson D, Bozkurt B, Deswal A, Mann DL, Spinale FG (2002) Plasma matrix metalloproteinase and inhibitor profiles in patients with heart failure. *J Card Fail* 8: 390–398

95   Fedak PWM, Altamentova SM, Weisel RD, Nili N, Ohno N, Verma S, Lee T-YJ, Kiani C, Mickle DAG, Strauss BH et al (2003) Matrix remodeling in experimental and human heart failure: a possible regulatory role for TIMP-3. *Am J Physiol Heart Circ Physiol* 284: H626–634

96   Corda S, Samuel JL, Rappaport L (2000) Extracellular matrix and growth factors during heart growth. *Heart Failure Rev* 5: 119–130

97   Sun Y, Weber KT (1998) Cardiac remodeling by fibrous tissue: role of local factors and circulating hormones. *Ann Med* 30: 3–8

98   Camelliti P, Borg TK, Kohl P (2005) Structural and functional characterisation of cardiac fibroblasts. *Cardiovasc Res* 65: 40–51

99   Hsueh WA, Law RE, Do YS (1998) Integrins, adhesion, and cardiac remodeling. *Hypertension* 31: 176–180

100 Brilla CG, Maisch B, Zhou G, Weber KT (1995) Hormonal regulation of cardiac fibroblast function. *Eur Heart J* 16: 45–50

101 Bujak M, Frangogiannis NG The role of TGF-[beta] signaling in myocardial infarction and cardiac remodeling. *Cardiovasc Res* 74(2): 184–195

102 Stawowy P, Margeta C, Kallisch H, Seidah NG, Chretien M, Fleck E, Graf K (2004) Regulation of matrix metalloproteinase MT1-MMP/MMP-2 in cardiac fibroblasts by TGF-beta1 involves furin-convertase. *Cardiovasc Res* 63: 87–97

103 Gurantz D, Cowling RT, Villarreal FJ, Greenberg BH (1999) Tumor Necrosis Factor-{alpha} upregulates angiotensin ii type 1 receptors on cardiac fibroblasts. *Circ Res* 85: 272–279

104 Kusuyama T, Yoshiyama M, Omura T, Nishiya D, Enomoto S, Matsumoto R, Izumi Y, Akioka K, Takeuchi K, Iwao H et al (2005) Angiotensin blockade inhibits osteopontin

expression in non-infarcted myocardium after myocardial infarction. *J Pharmacol Sci* 98: 283–289

105 Zhou X, Tan FK, Guo X, Wallis D, Milewicz DM, Xue S, Arnett FC (2005) Small interfering RNA inhibition of SPARC attenuates the profibrotic effect of transforming growth factor beta1 in cultured normal human fibroblasts. *Arthritis Rheum* 52: 257–261

106 Schellings MWM, Pinto YM, Heymans S (2004) Matricellular proteins in the heart: possible role during stress and remodeling. *Cardiovasc Res* 64: 24–31

107 Bradshaw AD, Puolakkainen P, Dasgupta J, Davidson JM, Wight TN, Helene Sage E (2003) SPARC-null mice display abnormalities in the dermis characterized by decreased collagen fibril diameter and reduced tensile strength. *J Invest Dermatol* 120: 949–955

108 Sasaki T, Gohring W, Mann K, Maurer P, Hohenester E, Knauper V, Murphy G, Timpl R (1997) Limited cleavage of extracellular matrix protein BM-40 by matrix metalloproteinases increases its affinity for collagens. *J Biol Chem* 272: 9237–9243

109 Sage EH, Reed M, Funk SE, Truong T, Steadele M, Puolakkainen P, Maurice DH, Bassuk JA (2003) Cleavage of the matricellular protein SPARC by matrix metalloproteinase 3 produces polypeptides that influence angiogenesis. *J Biol Chem* 278: 37849–37857

110 Cohn JN, Levine B, Olivari MT, Garberg V, Lura D, Francis GS, Simon AB, Rector T (1984) Plasma norepinephrine as a guide to prognosis in patients with chronic congestive heart failure. *N Engl J Med* 311: 819–823

111 Dixon IMC, Reid NL, Ju H (1997) Angiotensin II and TGF-β in the development of cardiac fibrosis, myocyte hypertrophy, and heart failure. *Heart Failure Reviews* 2: 107–116

112 Kirchengast M, Klaus M (1999) Endothelin-1 and endothelin receptor antagonists in cardiovascular remodeling. *Proc Soc Experimental Biol Med* 221: 312–325

113 Zannad F, Alla F, Dousset B, Perez A, Pitt B (2000) Limitation of excessive extracellular matrix turnover may contribute to survival benefit of spironolactone therapy in patients with congestive heart failure: insights from the randomized aldactone evaluation study (RALES). Rales Investigators. *Circulation* 102: 2700–2706

114 Rauchhaus M, Doehner W, Francis DP, Davos C, Kemp M, Liebenthal C, Niebauer J, Hooper J, Volk HD, Coats AJ et al (2000) Plasma cytokine parameters and mortality in patients with chronic heart failure. *Circulation* 102: 3060–3067

115 Fichtlscherer S, Rossig L, Breuer S, Vasa M, Dimmeler S, Zeiher AM (2001) Tumor necrosis factor antagonism with etanercept improves systemic endothelial vasoreactivity in patients with advanced heart failure. *Circulation* 104: 3023–3025

116 Tsuruda T, Costello-Boerrigter LC, Burnett JC Jr (2004) Matrix metalloproteinases: pathways of induction by bioactive molecules. *Heart Fail Rev* 9: 53–61

117 McMurray JJ (2004) Angiotensin inhibition in heart failure. *J Renin Angiotensin Aldosterone Syst* 5 (Suppl 1): S17–22

118 Opie LH, Commerford PJ, Gersh BJ, Pfeffer MA (2006) Controversies in ventricular remodelling. *Lancet* 367: 356–367

119 Pfeffer MA (1994) Mechanistic lessons from the SAVE Study. Survival and ventricular enlargement. *Am J Hypertens* 7: 106S–111S

120 Peng H, Carretero OA, Vuljaj N, Liao T-D, Motivala A, Peterson EL, Rhaleb N-E (2005) Angiotensin-converting enzyme inhibitors: a new mechanism of action. *Circulation* 112: 2436–2445

121 Tsuruda T, Boerrigter G, Huntley BK, Noser JA, Cataliotti A, Costello-Boerrigter LC, Chen HH, Burnett JC Jr (2002) Brain natriuretic peptide is produced in cardiac fibroblasts and induces matrix metalloproteinases. *Circ Res* 91: 1127–1134

122 Shimizu N, Yoshiyama M, Omura T, Hanatani A, Kim S, Takeuchi K, Iwao H, Yoshikawa J (1998) Activation of mitogen-activated protein kinases and activator protein-1 in myocardial infarction in rats. *Cardiovasc Res* 38: 116–124

123 Dhein S, Polontchouk L, Salameh A, Haefliger J-A (2002) Pharmacological modulation and differential regulation of the cardiac gap junction proteins connexin 43 and connexin 40. *Biology of the Cell* 94: 409–422

124 Bouzegrhane F, Thibault G (2002) Is angiotensin II a proliferative factor of cardiac fibroblasts? *Cardiovasc Res* 53: 304–312

125 Campbell SE, Janicki JS, Weber KT (1995) Temporal differences in fibroblast proliferation and phenotype expression in response to chronic administration of angiotensin ii or aldosterone. *J Mol Cell Cardiol* 27: 1545–1560

126 Sawyer DB, Siwik DA, Xiao L, Pimentel DR, Singh K, Colucci WS (2002) Role of oxidative stress in myocardial hypertrophy and failure. *J Mol Cell Cardiol* 34: 379–388

127 Li Y, Huang TT, Carlson EJ, Melov S, Ursell PC, Olson JL, Noble LJ, Yoshimura MP, Berger C, Chan PH et al (1995) Dilated cardiomyopathy and neonatal lethality in mutant mice lacking manganese superoxide dismutase. *Nat Genet* 11: 376–381

128 Sentman M-L, Granstrom M, Jakobson H, Reaume A, Basu S, Marklund SL (2006) Phenotypes of mice lacking extracellular superoxide dismutase and copper- and zinc-containing superoxide dismutase. *J Biol Chem* 281: 6904–6909

129 Siwik DA, Pagano PJ, Colucci WS (2001) Oxidative stress regulates collagen synthesis and matrix metalloproteinase activity in cardiac fibroblasts. *Am J Physiol Cell Physiol* 280: C53–60

130 Reunanen N, Li S-P, Ahonen M, Foschi M, Han J, Kahari V-M (2002) Activation of p38alpha MAPK enhances collagenase-1 (matrix metalloproteinase (MMP)-1) and stromelysin-1 (MMP-3) expression by mRNA stabilization. *J Biol Chem* 277: 32360–32368

131 Cohn JN (1997) Overview of the treatment of heart failure. *Am J Cardiol* 80: 2L–6L

132 Dalton WS, Friend SH (2006) Cancer biomarkers – an invitation to the table. *Science* 312: 1165–1168

133 Pfeffer MA (1998) ACE inhibitors in acute myocardial infarction. *Circulation* 97(22): 2192–2194

134 Pfeffer MA, Pfeffer JM, Steinberg C, Finn P (1985) Survival after an experimental myocardial infarction: beneficial effects of long-term therapy with captopril. *Circulation* 72: 406–412

135 Reinhardt D, Sigusch HH, Hensse J, Tyagi SC, Korfer R, Figulla HR (2002) Cardiac remodelling in end stage heart failure: upregulation of matrix metalloproteinase (MMP) irrespective of the underlying disease, and evidence for a direct inhibitory effect of ACE inhibitors on MMP. *Heart* 88: 525–530

136 Lindsey ML, Spinale FG (2005) Targeting matrix remodeling in cardiac hypertrophy and heart failure. *Drug Discovery Today: Therapeutic Strategies* 2: 253–258

137 Lindsey ML (2006) Novel strategies to delineate matrix metalloproteinase (MMP)-substrate relationships and identify targets to block MMP activity. *Mini Rev Med Chem* 6: 1243–1248

138 Peterson JT (2006) The importance of estimating the therapeutic index in the development of matrix metalloproteinase inhibitors. *Cardiovasc Res* 69: 677–687

Matrix metalloproteinases and inflammatory diseases of the central nervous system

Yvan Gasche and Jean-Christophe Copin

Department of Anesthesiology, Pharmacology and Intensive Care, Geneva University Hospitals and Geneva Neuroscience Center, University of Geneva, 1211 Geneva, Switzerland

Abstract

Matrix metalloproteinases (MMPs) are involved in the pathogenesis of several diseases of the central nervous system (CNS) that share common pathophysiological processes, such as blood-brain barrier (BBB) disruption, oxidative stress, remodelling of the extracellular matrix (ECM) and inflammation. In ischaemic brain injury, MMPs are implicated in various stages of the disease. MMPs contribute to the disruption of BBB leading to vasogenic oedema and to the influx of leucocytes into the CNS. The ability of MMPs to digest the basal lamina of capillaries increases the risk of haemorrhagic transformation (HT) of the ischaemic tissue. During the acute ischaemic phase, maintenance of the ECM is essential for neuronal survival. However, ECM degradation and its reconstitution are critical to tissue recovery. MMPs, as key modulators of ECM homeostasis, play a role in the cascades leading to neuronal cell death and tissue regeneration. Yet they may have a detrimental or beneficial role depending on the type and the stage of brain injury. This pleiotropic implication of MMPs in brain injury has opened new areas of investigation, which should lead to innovative therapeutic strategies.

Regulation of matrix metalloproteinases (MMPs) in the central nervous system (CNS)

MMPs belong to a family of Zn^{2+}- and Ca^{2+}-dependent endopeptidases including more than 20 members [1–3].

Most MMPs are structurally organised into three domains; an aminoterminal propeptide, a catalytic domain and a hemopexin-like domain at the carboxy-terminal. In most cases, MMPs are secreted in the interstitial space as an inactive zymogen. A subclass of MMPs is membrane-bound (MT-MMP) and features a transmembrane domain and a short cytoplasmic C-terminal end or a hydrophobic region with a specific function [4]. MMP-9 and MMP-2 show a specific gelatin-binding domain. Taking advantage of the specific affinity of MMP-9 and MMP-2 to gelatin, we and others have developed gelatin-sepharose based purification methods, to extract those specific MMPs from brain tissues [5, 6]. The introduction of these new techniques of extraction allowed high sensitivity measurements of MMP-9 and MMP-2 in brain, as compared to traditional methods. This technical innovation was at the origin of the discovery that MMP-9 was induced and activated very early after the onset of brain ischemia [5, 7].

Tissue expression of most MMPs is low under normal conditions and is induced when remodelling of ECM is necessary. MMP gene expression is mostly regulated at the transcriptional level, but modulation of mRNA stability has been described for stromelysin, collagenase and gelatinase A (MMP-2) in response to growth factors and cytokines [8, 9] and for gelatinase B (MMP-9) in response to nitric oxide (NO) [10]. The promoter gene sequence of several MMPs such as MMP-1, 3, 7, 9, 10, 12 and 13 contains an AP-1 site and an NF-κB site is identified on the promoter region of MMP-9 [11]. Stimulation or repression of these growth factor-cytokine-redox responsive MMP genes results in major changes in mRNA and protein levels. On the other hand, the human gelatinase A promoter does not contain any cytokine responsive element but only an AP-2 site and behaves as a housekeeping promoter [12], leading to an extensive expression of MMP-2 under normal circumstances, with a low responsiveness to inflammatory stimuli.

What should be kept in mind, regarding MMP regulation, is that cells do not indiscriminately release these enzymes. MMPs are secreted and anchored to cell membranes acting thereby on specific substrate located in the peri-cellular space. For example, gelatinase A binds to the integrin αvβ3 [13], gelatinase B to CD44 [14], and matrilysin to surface proteoglycans [15]. Since MMPs are secreted as latent enzymes, physiological or pathophysiological activation becomes a critical control point. The classic mechanism of MMP activation, 'the cysteine switch mechanism', consists of the disruption of the interaction between a zinc molecule on the active site and a cysteine in the pro-domain, leading to the auto-proteolytic cleavage of the zymogen and the production of the mature active form of the enzyme [16]. Chemical agents such as p-chloromercuribenzoate [17] or sodium dodecyl sulfate are able to generate active zymogen by exposing the zinc active site. Other agents such as reactive oxygen species or organomercurials inactivate the cysteine residue [18–20]. Early description of factors in conditioned cell culture medium activating human proMMP-1 with no modification of the enzyme molecular weight [21] suggested that non-proteolytic mechanisms of MMP activation could exist in biological systems. In addition, S-nitrosylation or N-glycosylation of gelatinase B can also modify its activity [22, 23]. Alternatively, proteolytic enzymes [17] may cleave the propeptide ahead of the cysteine, which will be removed after an autocatalytic digestion to produce a stable activated MMP. Among many proteases, plasmin and urokinase-type plasminogen activator and tissue-type plasminogen activator are known to serve as important physiological activators of MMPs [24]. Metalloproteinases may also interact with each other and further promote activation as it has been described for MMP-3 or MMP-2 activation of gelatinase B [4, 25]. In this regard, MMP-2 is activated by membrane-type MMPs (MT-MMPs). MT1-MMP, MT2-MMP, MT3-

MMP, and MT4-MMP are expressed at low levels in many cell types. Although, MT1-MMP, which is regulated by cytokines, is the most predominant MT-MMP [26], both MT2-MMP and MT3-MMP share with MT1-MMP the ability to initiate the activation of pro-MMP-2. In contrast, MT4-MMP has no ability to process proMMP-2 into its active form. In addition to the activation of proMMP-2, MT1-MMP is responsible for the activation of proMMP-13 [18]. Processing of proMMP-13 occurs *via* a 56 kDa intermediate and a final 48 kDa form. After activation, MMP catalytic activity is regulated by the tissue inhibitors of metalloproteinases (TIMPs). Four TIMPs are identified so far: TIMP-1, 2, 3 and 4. They interact with the active site plus a site in the carboxyl terminal hemopexin-like region of MMPs. The C-terminal region of TIMPs interacts with the C-terminal region of the enzyme, increasing the rate of association many fold. TIMP-1, the natural inhibitor of MMP-9, prevents pro-MMP-9 activation. Furthermore, the pro-MMP-9–TIMP-1 complex can interact with MMP-3 and then dissociate into free pro-MMP-9 and the TIMP-1–MMP-3 complex. TIMP-2, the natural inhibitor of MMP-2, plays a dual role by interacting with this gelatinase. Indeed, when TIMP-2 complexes with proMMP-2, cell surface-mediated activation of the enzyme is facilitated, whereas interaction of TIMP-2 with the active enzyme results in its inhibition [27–31].

Regulation of MMP expression and activity appears to be a very complex and tightly controlled process that is not yet fully unravelled, especially *in vivo*.

MMP expression in ischaemic brain injury and other CNS pathologies

The expression of MMPs is very low in the normal adult brain. MMP expression varies depending on the different brain regions, cell populations and state of development, as recently reviewed by Dzwonek et al. [32]. MMP-2 was identified in the cortex, cerebellum, hippocampus, mainly in astrocytes, but also in some neurons such as Purkinje cells or cortical neurons [33–36]. MMP-9 was found in the cortex, cerebellum, and the hippocampus, while MMP-3 was present in the cerebellum [33–36]. TIMPs were also found under normal conditions in different brain regions and cell types [32].

MMP expression in ischaemic brain injury

Experimental *in vivo* studies, investigating the role of MMPs in stroke, have shown that gelatinase expression was induced during experimental focal ischaemia. Studies, conducted in spontaneously hypertensive rats resulted in some skepticism regarding the role of MMPs in the early phase of ischaemic brain injury, since these studies could not show any upregulation or activation of the enzymes during the early hours after permanent middle cerebral artery occlusion (MCAO) [37, 38]. Later studies, using permanent or transient models of MCAO and more sophisticated techniques of MMP extraction showed, however, that gelatinase B was already upregulated 1 to 2 h after ischaemia [5, 39]. We could show that gelatinase B upregulation was rapidly followed by the appearance of the active form of the enzyme [5, 7]. One study, conducted in baboons, showed an early upregulation of gelatinase A [40]. As a corollary to these experimental studies, in human ischaemic stroke, MMP-9 expression was shown to be mainly increased in acute ischaemic lesions (less than 1 week after stroke onset) while MMP-2 and MMP-7 (matrilysin) were observed in chronic lesions (more than 1 week after stroke onset) [41, 42].

Free radical formation is an early and central event in the pathophysiology of brain ischaemia-reperfusion [43]. We could observe that the oxidative unbalance induced by ischaemia-reperfusion or mechanical brain injury is a key event in MMP 9 overexpression [44, 45]. As mentioned earlier, MMPs are expressed at low levels in resident cells of the normal brain. Under pathological conditions, the exact cellular source of the MMPs involved in BBB disruption is not known. Recently, Gidday et al. (St Louis, MO, USA) [46] created chimeric MMP-9 knockout animals producing MMP-9$^{+/+}$ leucocytes or wild type animals producing MMP-9$^{-/-}$ leucocytes in order to address this issue. Subjecting these animals to stroke allowed him to identify recruited leucocytes, as a key cellular source of cerebral MMP-9 in the early phase following experimental focal ischaemia.

Earlier studies [37] examined the expression of MMP-1 and MMP-3 (by western blotting) after permanent MCAO. These studies failed to show any expression of these proteins. Rosenberg et al. [47] could not find any MMP-3 immunostaining in normal cerebral cortex of spontaneously hypertensive rats, but they observed positive immunoreactivity for MMP-3 in activated microglia and ischaemic neurons, 24 h after reperfusion, using a model of transient MCAO [47]. Inconsistent results were also found for TIMP-1 expression. While some studies showed an overexpression of TIMP-1 at the mRNA and protein level starting at 12 h after the ischaemic insult and reaching a peak at 48 h [48, 49], we and others could not find any modulation of TIMP-1 expression during the first 24 h [5, 37]. On the other hand, Nguyen et al. [50] demonstrated that TIMP-1 independent activated MMP-9 was produced in endothelial cells, implying a more intricate intracellular regulation of the enzyme than previously thought.

The inconsistent results of the expression of MMPs and TIMPs in brain ischaemia may be related to methodological particularities of the experimental models, to variations in MMP measurement techniques and to species related specificities.

MMP expression in other CNS pathologies

In multiple sclerosis (MS), immunohistochemistry of brain tissue demonstrated that the expression of MMP-1, 2, 3, 7, 9 by astrocytes, microglia, endothelial cells and non resident cells such as T cells and macrophages, was increased in and around plaques [41, 51, 52]. During the development of experimental autoimmune encephalomyelitis (EAE), the increase of MMP-7 and 9 in brain tissue peaked at the time when clinical symptoms became evident [53]. In human, MMP-9 levels were increased in cerebrospinal fluid (CSF), whereas MMP-2, 3 and 7 levels were unaltered [54, 55]. While MMP-9 CSF levels were similarly increased in MS patients with acute relapses and those in clinically stable phases of disease [55], MMP-9 serum levels appeared to increase during clinical relapses and high levels correlating with an increased number of gadolinium-enhancing MRI lesions [56] or with the risk of developing new gadolinium-enhancing lesions [57]. Several studies have consistently shown that serum levels of MMP-9 were significantly higher in MS patients, compared to healthy control subjects [56–58]. Accordingly, MMP-9 protein levels in peripheral blood mononuclear cells of active MS were significantly higher compared to controls [59]. Furthermore, TIMP-1 was found to be upregulated in chronic plaques [60], while concentrations of this inhibitor were low in CSF and plasma of MS patients [56, 59, 61–63]; the level of TIMP-1 increased after treatment with interferon-beta (IFN-β).

Bacterial meningitis is another example of inflammatory disease, where infectious organisms trigger a host inflammatory response. It is not surprising that upregulation of MMPs was consistently observed in the course of bacterial meningitis [54, 64–66]. MMP-9 and MMP-8 were shown to be upregulated in the CSF from children with bacterial meningitis, whereas MMP-2, 3 were not present [67]. We recently reported a dramatic increase of MMP-9 in the CSF of a meningitic patient developing parainfectious vasculitis [66]. Accordingly, CSF levels of MMP-9 were shown to be significantly higher in patients who developed neurological sequelae than in those who recovered [67]. Experimental rat models of pneumococcal meningitis showed a dramatic induction of MMP-3, 8, 9, 12, 13, and 14, but not MMP-2 and 7, in brain parenchymal tissue. In cells found in the CSF, MMP-8 and 9 mRNA were significantly increased, while MMP-2 and 7 mRNA remained at basal levels [68, 69].

Overexpression of MMP-2 and MMP-9 has also been noted following traumatic and viral brain injury, as well as after kainate treatment [36, 70–72]. In the course of kainate-induced seizure, early expression of MMP-9 (6–24 h) was seen in the hippocampal dentate gyrus and neocortex. MMP-9 expression was observed in neuronal cell bodies as well as in dendritic layers [73]. Several studies also showed that not only MMP-9 was increased, in the hippocampus following seizures, but TIMP-1 levels were increased as well [74, 75].

Role of MMPs in ischaemic brain injury and other CNS pathologies

MMPs are upregulated in most of CNS pathologies inducing an inflammatory response. Since MMPs are not only key modulators of the ECM, but also critical effectors for the cleavage or shedding of various molecules (to their active or inactive forms) including pro-inflammatory cytokines such as interleukin-1 beta (IL-1β) [76] and tumour necrosis alpha (TNF-α) that are implicated in many inflammatory processes, MMPs appear to be involved at many different pathophysiologic levels in various brain pathologies.

MMPs in ischaemic brain injury and disruption of the blood brain barrier

The main causes of early neurological deterioration and mortality following ischaemic stroke are brain oedema [77] and HT [78]. The common pathophysiological pathway leading to these complications is the disruption of the BBB. The role of BBB is to preserve the neuronal microenvironment, which is essential for cell survival and the normal function of the brain. Several groups, including ours, have demonstrated that BBB disruption is taking place in ischaemic stroke [5, 79–81] already a very few hours following the ischaemic event. BBB disruption leads to vasogenic oedema formation.

Ischaemia also rapidly triggers a pro-inflammatory cascade at the blood-vascular-parenchymal interface [82] that is further exacerbated by reperfusion [83]. Free radical formation is an early and central event in the pathophysiology of brain ischaemia-reperfusion [43]. Among other stimuli, oxidative stress triggers pro-inflammatory cytokines such as TNF-α and IL-1β, which were increased within a few hours of the insult [84]. Accordingly, post-reperfusion treatment with anti-TNF-α neutralising antibody reduced brain infarct volume and cerebral oedema, as well as cerebral MMP-9 overexpression in experimental stroke [85]. These pro-inflammatory events signal vascular and leucocyte activation, leading to the appearance or increase of adhesion molecules for leucocytes on the surface of microvascular endothelial cells [86, 87] followed by endothelial-leucocyte adhesion and finally leucocyte penetration across the BBB into the ischaemic tissue. Transmigrated leucocytes will further promote brain injury by releasing oxygen free radicals and various proteolytic enzymes. As a functional entity, BBB includes several cell types and the ECM. The endothelial basal lamina represents the non-cellular component of the BBB. Endothelial cells synthesise basement membrane proteins; the basal lamina is a specialised ECM composed of type IV collagen, fibronectin, laminin and various proteoglycans [88]. The ECM components interact with endothelial cells *via* integrins and regulate distinct biological events such as cellular differentiation, survival, morphology, adhesion and gene expression [89–92]. During cerebral ischaemia-reperfusion, the ECM is disrupted. Major components of the endothelial basal lamina such as laminin, type IV collagen and fibronectin start to disappear as soon as 2 h after the onset of ischaemia [93]. The first signs of BBB leakage were consistently observed between 2 to 8 h after the onset of ischaemia [5, 48, 94, 95]. By 24 h of experimental ischaemia-reperfusion, dissolution of microvascular structures led to clear evidence of microvessel interruption [93, 96, 97] with local haemorrhage [96, 97]. Thus, proteolysis seems to be a critical process of stroke related to BBB disruption.

All of the endothelial basal lamina components can be digested by MMPs. Among the MMPs, gelatinase A and gelatinase B specifically digest type IV collagen in the basal lamina. Rosenberg et al. [48] confirmed the role of MMPs in BBB disruption by demonstrating the ability of the non-selective MMP inhibitor BB-1101 (British Biotechnology, UK) to reduce early BBB leakage following transient focal ischaemia. Asahi et al. [98] used knockout mice to demonstrate that among the MMPs, MMP-9 was the main MMP contributing to BBB disruption in experimental focal cerebral ischaemia, while MMP-2 has no influence on brain injury [99]. Recently, Fukuda et al. [100] formally demonstrated the role of activated proteases in vascular degradation.

The possible mechanism accounting for MMP mediated BBB alteration is the disruption of the endothelial basal lamina that prevents the anchorage of the endothelial cells onto the ECM. MMP-9 may also induce the loosening of the tight junctions by the alteration of their constituents, such as zonula occludens protein-1 [98] and other proteins such as claudin-5 [101]. MMPs indirectly affect BBB permeability by interfering with inflammatory pathways triggered by ischaemia-reperfusion. From *in vitro* studies, we know that MMPs can process IL-1β into its biologically active form. While MMP-2 activates proIL-1β in 24 h, MMP-3 takes 1 h and MMP-9 only a few minutes to process the proIL-1β into its active form. Similar observations were made for IL-8 [102]. Thus, MMPs may promote inflammatory processes, in a positive feedback loop increasing MMP production by resident or migrating cells.

The fundamental role played by MMPs in the development of vasogenic oedema during stroke is further substantiated by the fact that these enzymes mediate the capillary leakage triggered by oxidative stress, as demonstrated by investigations we conducted a few years ago [44]. It is well established that the oxidative unbalance during focal cerebral ischaemia is a major contributor to BBB disruption, secondary brain injury and HT transformation [95, 103–105]. Reactive oxygen and reactive nitrogen species are generated through several different cellular pathways, including calcium activation of phospholipases, nitric oxide synthase, xanthine oxidase, the Fenton and Haber-Weiss reactions, in various inflammatory and non-inflammatory cells. During ischaemia-reperfusion the burst of free radicals may overwhelm cellular defenses and lead to oxidation of lipids, proteins, and nucleic acids, which may alter cellular function in a critical way. Recently, genetic manipulation of wild-type animals has yielded species that over- or under-express genes such as, copper-zinc superoxide dismutase (CuZnSOD, SOD1) and manganese superoxide dismutase (MnSOD, SOD2). The introduction of the species has improved the understanding of oxidative stress. During focal ischaemia-reperfusion, mice deficient in CuZnSOD present with a more serious vasogenic oedema than their wild type counterparts [95]. We observed, in these CuZnSOD deficient animals, a rapid overproduction

of MMP-9 in the ischaemic tissue, after MCAO. The MMP-mediated *in situ* proteolysis, that we detected at the ischaemic capillary level, correlated with the local
production of oxygen free radicals. We finally could observe that vasogenic oedema
was dramatically reduced in these oxidative stress-sensitive animals by the inhibition of MMPs [44]. Similar results were observed in animals deficient in MnSOD
[106]. These animals, sensitised to oxidative stress by a 50% reduction in the mitochondrial form of SOD, showed a higher rate of reperfusion-related brain haemorrhage, larger infarct volumes and cerebral vasogenic oedema, as well as higher level
of MMP-9 in the ischaemic tissue than wild type littermates.

Microvascular disruption due to MMP proteolytic activity is a key event in ischaemic stroke. This suggests that MMPs may be responsible for the increased risk
of HT following reperfusion with rtPA. Recent experimental data confirmed this
hypothesis. In a model of thromboembolic stroke followed by reperfusion with rtPA
in the rabbit, Lapchak et al. [107] showed that a non-selective inhibition of MMPs,
before the onset of ischemia, was able to reduce the HT. However, it is still unclear
whether MMP inhibition during reperfusion after the onset of ischaemia would
efficiently reduce BBB damage and HT. Pfefferkorn et al. [108] showed a reduction
in mortality of rats subjected to MCAO followed by reperfusion with rtPA and
treated with an inhibitor of MMPs 2 h after the onset of the ischaemic event. In this
study however, based on a non-embolic model of MCAO, MCA recanalisation did
not provoke any intracerebral haemorrhage [108]. The lack of the haemorrhagic
component and the non-embolic nature of MCA occlusion in this model prevent a
straightforward translation of the data. The authors suggested that the preservation
of BBB reduced rtPA diffusion into the brain parenchyma and reduced the potential
exitotoxic effect of the thrombolytic agent [108]. Thus far, it is not known whether
a delayed MMP inhibition would extend the therapeutic window for thrombolysis
by efficiently reducing the risk of HT and brain oedema.

In order to address this issue, we investigated, in a large group of Sprague-Daw-
ley rats (215) subjected to 6 h thrombo-embolic MCAO followed by reperfusion
with rtPA, the effect of a 3 to 6 h delayed MMP-inhibitor treatment on the development of brain oedema, HT and neurological impairment (unpublished observation). We first confirmed that intraperitoneal injection of the broadspectrum MMP
inhibitor before MCAO significantly reduced the risk of unfavourable neurological
outcome at 24 h. Neurological protection was slightly weakened by postponing the
MMP inhibitor treatment for 3 h after the onset of ischaemia. Pre and early post-
ischaemic treatment resulted in a slight but not significant reduction of mortality at
24 h. Late post-treatment at 6 h after occlusion did not improve the neurological
score and did not reduce mortality. While pre-treating the animals with the MMP
inhibitor before MCAO dramatically decreased the risk of severe haemorrhagic
transformation from 60% to 13%, postponing MMP inhibition for 3 or 6 h after
MCAO increased the number of animals that bled. Nevertheless, a 3 or 6-h delay
in MMP inhibition still significantly reduced the risk of HT as compared with

vehicle-treated animals. In vehicle-treated rats, 6-h ischemia induced brain swelling characterised by a hemispheric enlargement of 24%, 24 h after MCAO. The MMP inhibitor, injected immediately before or 3 h after the onset of ischaemia, significantly reduced cerebral oedema. But delaying MMP inhibition for 6 h after MCAO did not decrease cerebral oedema as compared with control animals.

Our study showed for the first time, in an embolic model of stroke where thrombolysis with rtPA was administered beyond the 3 h therapeutic window, that the risk and gravity of HT could be reduced by injecting a broad spectrum MMP inhibitor after the onset of ischaemia, at a time point that is clinically relevant.

Finally, in this study, we have confirmed experimentally that the risk and intensity of haemorrhagic transformation after thrombolysis with rtPA, as well as the degree of neurological impairment, increased with the duration of MCAO. Hence, more than half of the animals showed signs of haemorrhagic infarction or more severe parenchymal haemorrhage when the period of ischemia was longer than 3 h. For occlusions exceeding 6 h, the quantity of haemoglobin released in the ischaemic parenchyma doubled and almost half of the animals died within 24 h. However, we noticed many cases of partial occlusions after clot injection and many cases of spontaneous reperfusion. Consequently, among the 215 rats that were operated, more than 134 were not considered in the final analysis due to improper artery occlusion or recanalisation. Our data, in agreement with studies published by others [109, 110] point out that thromboembolic models show significant variability in terms of artery occlusion and recanalisation, and emphasise the need to periodically assess regional cerebral perfusion in models evaluating therapeutic strategies in order to avoid questionable interpretations of the outcomes.

Thus, accumulating data show that the alteration of MMP regulation is a major contributor to HT after reperfusion. In addition, recent studies suggest a close interaction between MMP and rtPA pathways. Indeed, while exogenous rtPA increased the ischaemic level of MMP-9 in experimental stroke, rtPA knockout mice showed decreased levels of MMP-9 [111, 112]. On the other hand, it was also shown that an increase in endogenous rtPA activity in the perivascular tissue following cerebral ischaemia induced opening of the BBB *via* a mechanism that was independent of both plasminogen and MMP-9 [113]. By contrast, Wang et al. [114] observed that MMP-9 played a detrimental role in a murine model of brain haemorrhage unrelated to ischaemia.

Inhibition of MMP-9 protects the adult brain after cerebral ischaemia. However, until recently the role of MMP-9 in the immature brain after hypoxia–ischaemia was unknown. In a model of post-natal hypoxia-ischaemia Svedin et al. could observe a delayed and diminished leakage of BBB and a decrease in inflammation in MMP-9-deficient mice as compared to wild type animals. This protection was not linked to either caspase-dependent or caspase-independent apoptotic processes [115].

One should underscore that clinical investigations resulted in observations that are in accordance with those made in experimental studies. Thus, in human ischaemic stroke, basal MMP-9 plasma levels obtained before rtPA treatment, appear to predict intracranial haemorrhagic complications occurring after thrombolysis [116] with a good sensitivity and specificity [117]. On the other hand, a positive correlation was found between MMP-9 plasma levels and the infarct size [118, 119]. So far, however, it is not known whether measuring MMP-9 plasma levels at the admission in stroke patients would help to determine those who could be thrombolised safely outside of the 3 h therapeutic window and those who should not be thrombolised despite being within the therapeutic window.

An additional point that was raised in an attempt to unravel the mechanism of BBB disruption and haemorrhagic transformation was the issue of identifying the precise cellular source of gelatinase implicated in vascular proteolysis. Although neurons [120], astrocytes [121], microglial cells [122], endothelial cells [37] and oligodendrocytes [123], express gelatinases, recent data suggested that MMP-9 secreted by transmigrated leucocytes was a major contributor to BBB disruption. As mentioned earlier, Gidday et al. [46] used MMP-9 knockout mice and chimeric knockout animals lacking MMP-9 either in leucocytes or in resident brain cell to determine if MMP-9 released from leucocytes and recruited into the brain during post-ischaemic reperfusion, contributed to BBB disruption and brain injury. In this study, one [46] observed that the extent of BBB breakdown, the neurologic deficit, and the volume of infarction after transient focal ischaemia were abrogated to a similar extent in MMP-9 knockout mice and in chimeras lacking leucocytic MMP-9, but not in chimeras with MMP-9-containing leucocytes. Zymography and western blot analysis of MMP-9 from these chimeras confirmed that the elevated MMP-9 expression in brain at 24 h following reperfusion largely derived from leucocytes. Interestingly, MMP-9 knockout mice showed a reduction in leucocyte-endothelial adherence and in the number of neutrophils plugging capillaries as well as infiltrating the ischaemic brain during reperfusion. Microvessel immunoreactive collagen IV was also preserved in these animals [46].

We further demonstrated the blood-born origin of MMP-9 by showing that *in situ* MMP inhibition by intracerebrovascular injection of hydroxamate MMP inhibitors was unable to prevent vasogenic oedema after transient MCAO, while the inhibiton of MMPs by the traditional intraperitoneal route was effective in reducing BBB damage [81].

Moreover, we observed that depleting circulating leucocytes in mice before stroke not only dramatically reduced the induction of MMP-9 observed in the ischaemic hemisphere but also significantly reduced the risk of HT after thrombolysis (unpublished observation).

All together, the results of these experiments re-emphasised the role of leucocyte-derived MMPs in ischaemia-related brain injury. The pathophysiological complexity of leucocyte involvement in stroke combined with methodological weaknesses in other studies probably explains the conflicting results obtained in human studies that examined the efficacy of anti-leucocyte strategies [124, 125]. Finally, cortical spreading depression, which is a propagating wave of neuronal and glial depolarisation implicated in stroke, was also shown to alter BBB permeability by activating MMP-9 [126].

MMPs in ischaemic brain injury and neuronal apoptosis

In response to internal or external death stimuli, mitochondria and other organelles may initiate apoptosis through the release of cytochrome c and activation of the intrinsic caspase pathway [127, 128]. Conversely, mitochondria may release apoptosis-inducing factor and initiate apoptosis by caspase-independent mechanisms [129, 130]. Activation of cell surface receptors including Fas and TNF-α receptor can also initiate apoptosis through the activation of caspase-8, which can activate the extrinsic caspase pathway by cleaving the proapototic molecule Bid that translocates to the mitochondria resulting in the release of cytochrome c [131, 132]. Finally, oxidative DNA damage can trigger apoptosis by activating the transcription factor p53, upregulating transcription of the Bax gene that encodes a pro-apoptotic protein with mitochondrial membrane permeabilisation-inducing properties [133]. Caspase-3 is the major downstream apoptosis effector enzyme. However, calpain, another cysteine protease, also appears to play a critical role in apoptosis, as judged by the anti-apoptotic effect of calpain inhibitors [134]. Despite differences in cleavage-site specificity, both proteases cleave many common substrates, such as poly (ADP) ribose polymerase (PARP) and alpha-spectrin.

Brain ischaemia-reperfusion triggers various pathophysiologic cascades, which interact in a global network with positive feedback loops leading to neuronal cell death. The major participants to this network are oxidative stress and pro-inflammatory cytokines. In this context, several experimental studies have shown that strategies designed to reduce oxygen free radical formation [43] or pro-inflammatory cytokines [135] such as TNF-α and IL-1β release were efficacious in limiting neuronal injury. Oxidative stress was shown to promote the mitochondria-dependent (i.e., intrinsic) apoptosis pathway which is a major mechanism resulting in neuronal apoptosis in ischaemia [136]. MMPs, including MMP-2, MMP-3 and MMP-9, can convert the inactive precursor form of IL-1β into biologically active forms, which are implicated in the development of brain damage following cerebral ischaemia. MMP involvement in delayed cell death may also be related to the disruption of BBB leading to brain oedema and secondary cell injury, but the recent observation that neuronal survival is closely related to the maintenance of the ECM [90], suggested a more direct role for MMPs in the pathophysiology of ischaemic neuronal injury. Gu et al. [22] demonstrated that MMP-9 activation by S-nitrosylation induced neuronal apoptosis *in vitro*. During cerebral ischaemia *in vivo*, MMP-9 co-localised with neuronal nitric oxide synthase. Activated MMP-9 was identified, both *in vitro* and *in vivo*, as a stable sulfinic or sulfonic acid, whose formation was triggered by

S-nitrosylation [22]. Thus far only global ischaemia and kainate models of brain injury have investigated the implication of MMPs in neuronal apoptosis *in vivo* or *ex vivo* [137, 138]. In that regard, Lee et al. [138] showed that selective neuronal cell death in the hippocampus was reduced in MMP-9-deficient mice and also in animals treated with a non-selective inhibitor of MMPs when compared to control animals. However, the exact mechanism leading to the protection was not investigated. Jourquin et al. [137] also showed that non-selective inhibition of MMPs, as well as selective inhibition of MMP-9 was capable of protecting vulnerable neurons after kainate challenge to organotypic neuronal cultures.

Recent investigations conducted in our laboratory showed that non-selective MMP inhibition reduced cerebral infarct as well as DNA fragmentation after experimental focal ischaemia. The cerebroprotective effect occurred concomitantly with a reduction of cytochrome c release into the cytosol, a reduction of calpain-related α-spectrin degradation, as well as an increase in the immunoreactivity of the intact form of PARP [81]. By contrast, specific targeting of the *mmp-9* gene in mice did not modify the apoptotic response after cerebral ischaemia, although the intra-cerebroventricular injection of a non-selective inhibitor of MMPs in MMP-9 deficient mice provided a significant reduction of DNA degradation. These results indicated that MMPs other than MMP-9 are actively involved in cerebral ischaemia-induced apoptosis.

The involvement of TIMPs as regulators of apoptosis was also studied. Glutamate-induced excitotoxicity was attenuated by TIMP-1 in cultured neurons [139]. Interestingly, cytoprotection seemed to be independent of MMPs, since the nonselective inhibition of MMPs was unable to reproduce the cytoprotective effect produced by TIMP-1. TIMP-3 may also play a crucial role in ischaemic neuronal cell death. Indeed, TIMP-3 may stabilize the interaction between TNF-α and its receptor as well as between FasL and Fas through the inhibition of MMPs since MMPs have protein-shedding abilities. While TIMP-3 is expressed at very low levels in normal brain tissue, TIMP-3 is highly expressed in ischaemic cortical neurons undergoing apoptosis after experimental MCAO [47, 140]. In addition, TIMP-3 and MMP-3 modified neuronal sensitivity to Fas-mediated apoptosis induced by doxorubicin [141]. In this study, MMP inhibition by TIMP-3 appeared to be necessary for doxorubicin-induced apoptosis. MMP-3 added to cell cultures markedly attenuated apoptosis and blunted Fas receptor–FasL interactions at the neuronal cell surface [141].

MMPs in ischaemic brain injury, astrogliosis and long-term recovery

Whenever the nervous system is damaged it undergoes an injury response, so-called reactive gliosis or glial scarring, that evolves over several days [142]. The main cell types involved in these changes are astrocytes, macrophages, microglial cells and oligodendrocyte precursors [143–145]. Among these cells, the first ones to arrive at the location of injury are white blood cells from the bloodstream and microglia migrating in from the surrounding tissue. After 3–5 days, large numbers of oligodendrocyte precursor cells are recruited from the surrounding tissue. The final structure of the glial scar is predominantly astrocytic, and these cells divide and slowly migrate from the adjacent undamaged parenchyma towards the injured site, eventually to fill in the vacant space [142]. The consequence is that after the initial period of cell damage and death, during which neuroprotective treatments are the main therapeutic aim, any form of treatment designed to repair CNS injury, whether it be designed to make axon regenerate or to replace dead neurons, will inevitably have to take place in a glial scar environment. A major cause of failure of axon to regenerate in the CNS is the inhibitory nature of such an environment [146]. Nevertheless, the astrocytic component of the glial scar may also contribute actively to the repair of BBB after brain injury (Bush et al., 1999). Recently some attention has been paid to the role of MMPs in cell migration and CNS regeneration [147]. The contribution of MMPs in facilitation of cell migration has become evident [148]. In the nervous system, oligodendrocytes utilise MMP-9 to extend their process [123]. MMP-2 and MMP-9 promote axonal outgrowth in injured peripheral nerve [149] and are transiently but vigorously induced during maturation of the scar tissue of injured adult spinal cord [149, 150]. MMPs could also play a role in the formation of the glial scar by potentiating microglial osteopontin and thus facilitating reactive astrocyte recruitment [151, 152]. In a recent study [153], we documented the effect of MMP-9 depletion on glial scar formation in order to potentially extent the beneficial therapeutic role of MMP-9 inactivation during cerebral ischaemia to post-ischaemic processes, cellular recruitment and tissue remodelling, or to rule out potential long lasting side effects with unknown consequences of MMP inhibition. In a model of focal ischaemia-reperfusion carried out in mmp9$^{-/-}$ and wild-type mice, we analysed the glial scar formation within infarct areas, which were positively stained with Fluoro-jade on adjacent sections, in order to avoid a bias due to the reduction of the infarct size. 45 mins of MCAO induced neuronal degeneration in the same cerebral areas in wild-type and mmp9$^{-/-}$ mice. An intense neuronal degeneration was detected after 24 h within the ipsilateral striatum and a more sporadic degeneration was seen within the ipsilateral cortex. The count of Fluoro-jade positive cells at 1 day of recovery gave an average of 1525 ± 145 cells/mm^2 (wild type) and 1422 ± 330 cells/mm^2 (mmp9$^{-/-}$) in the striatum and 1460 ± 130 cells/mm^2 (wild type) and 1277 ± 444 cells/mm^2 (mmp9$^{-/-}$) in the cortex. Those values were not statistically different. Staining of degenerating neurons after 4 days of recovery was very similar to what we observed after 1 day of recovery. On day 10, degenerating neurons were preferentially localised at the edge of the ischaemic core in the striatum and in narrower cortical areas. After 21 days of recovery, degenerating neurons were barely detectable.

24 h after ischaemia, reactive astrocytes were almost undetectable in any of the two genotypes and immunostainings were not significantly different between the

ipsilateral and the contralateral striatum. Positive astrocytes showed a delicate staining of their processes as indication of their latency. At 4 days of recovery, the striatal infarct was surrounded by a large territory of reactive astrocytes both in wild-type and mmp9$^{-/-}$ animals. Reactive astrocytes had a typical pattern showing fat cell bodies with thick processes. In the cortex, reactive astrocytes were seen spreading homogeneously within the injured areas. Those cells were not seen in the contralateral cortex. The density of reactive astrocytes was identical between wild-type and mmp9$^{-/-}$ animals. On day 10, reactive astrocytes were seen within the entire striatal infarct, both in wild-type and mmp9$^{-/-}$ mice. Immunostainings done after 21 days of recovery were similar to those done after 10 days of recovery, regardless of the genotypes, and demonstrated the long-lasting astrocyte reactivity due to temporary MCAO both in the striatum and the cortex. The kinetic of microglial cell and oligodendrocyte precursor activation was also similar in wild type and knockout animals over a 3-week period. Only a slight difference in the pattern of macrophage infiltration was observed. These results suggest that a specific targeting of MMP-9, as a mean to prevent ischaemia-induced blood brain barrier disruption, would have no significant effects on the recruitment of cells involved in glial scar formation. Based on our current findings, we suggest that preservation of the BBB by MMP-9 inactivation should have no impact on the recruitment of cells involved in glial scar formation. We cannot however rule out the possibility that mmp9$^{-/-}$ animals have undergone compensatory mechanisms that mask the effects of MMP-9 inactivation. Additionally, it must be noticed that most of the available MMP inhibitors have a broad-spectrum action that makes them inappropriate for the specific and exclusive targeting of MMP-9. Nevertheless, MMPs may also play more subtle roles at the molecular level and may be beneficial during neuronal regenerations by degrading antagonistic molecules within the glial scar.

Most of the experimental studies conducted to evaluate the role of MMPs in ischaemic brain injury suggest that controlling the proteolytic cascade during the acute phase of injury would by neuroprotective. However, the precise long-term consequences of MMP inhibition on the neurological prognosis and recovery are not known.

Zhao et al. observed that MMPs play a role in delayed cortical responses after focal ischaemia in rats [154]. MMP-9 was upregulated in the peri-infarct cortex after one week and co-localised with markers of neurovascular remodelling. A delayed inhibition of MMPs was responsible for an impairement of recovery.

Rosenberg et al. also suggested in their experimental study carried out in rats that the early inhibition of MMPs by a broadspectrum inhibitor could have long-term detrimental effects [155], despite an efficient reduction of BBB disruption obtained during the acute phase of stroke.

Still, an open-label, evaluator-blinded clinical study in stroke patients, showed that Bartel Index scores were significantly higher in minocycline-treated patients [156]. The tetracycline was administered orally for 5 days, with a therapeutic window of 6 to 24 h after onset of stroke. Interestingly haemorrhagic transformation was not different between groups. Tetracyclines are known to inhibit MMPs, a wide range of pro-inflammatory cytokines and processes, and to modulate the apoptotic cascade [157, 158].

MMPs in other CNS pathologies

The importance of BBB disruption and MMP implication in ischaemic stroke is well established. In haemorrhagic stroke more conflicting results were obtained regarding the specific role of MMPs in the process of spontaneous intra-parenchymal bleeding. Nevertheless therapeutic strategies aiming at reducing the local inflammatory response implicating MMPs and cytokines could improve haemorrhagic stroke prognosis [159].

MS is an autoimmune disease characterised by demyelination and axonal loss. MMPs disrupt myelin [160] and fragments of the MMP mediated digestion of myelin basic protein (MBP) after injection into rodents can induce EAE, the experimental model of MS [161, 162]. Also, human MMP-9 cleaves human myelin basic protein into peptide fragments, one of which contains the immunodominant epitope [162]. Non-selective inhibitors of MMPs improved or prevented EAE [1]. Young mice lacking MMP-9 have less symptoms after the induction of EAE when compared with wild type animals [163]. In humans, IFN-β inhibits the production of MMP-9 by leucocytes *in vitro*, and alters the capacity of leucocytes to cross endothelial [164] and ECM barriers [165, 166]. In addition, MS patients treated with IFN-β showed a decrease in serum MMP-9 as well as the number of leucocytes producing the proteinase [167]. Interestingly, activated MMP-9 is capable of degrading and inactivating IFN-β [168]. IFN-β degradation can be inhibited by minocycline [168].

Other MMPs including MMP-12 also appear to play a critical role in MS, as macrophages from MMP-12 deficient mice showed a diminished capacity to penetrate basement membranes *in vitro* and *in vivo* [169]. As in the case of ischaemic brain injury the ability of MMPs to shed critical pro-inflammatory cytokines, adhesion molecules, TNF-α and Fas receptors is a potential mechanism implicating MMP in MS pathophysiology.

Despite progress in intensive care and effective antimicrobial chemotherapy, bacterial meningitis is still associated with a high mortality and incidence of neurological sequelae [170–174]. Following pneumococcal autolysis, a rapid increase of pro-inflammatory cytokines (IL-1β, TNF-α, IL-6) and chemokines (IL-8, MIP1-2) is detected in CSF, followed by increased BBB permeability [175]. Over the past few years, experimental studies provided converging evidence for a central role of MMPs in bacterial meningitis [67–69, 176–181]. Indeed, Leib et al. [68, 69] showed that an association between a broad-spectrum inhibitor of MMPs and antibiotics reduced neuronal necrosis and apoptosis in a model of experimental meningitis.

The important results of these studies were a significant reduction in mortality, seizure incidence and a preservation of learning capacity in animals treated with the MMP inhibitor. The results of the studies by Leib et al. [68, 69] were, however, challenged by Bottcher et al. [182] who showed that MMP-9-deficient mice infected with *Streptococcus pneumoniae* were not protected when compared to wild type animals, probably due to a delayed bacterial blood clearance. This result could be explained by the fact that the animals in this study did not receive any antibiotic following bacterial CNS infection. Most recently, the potential involvement of MMPs in meningitis was further supported by Pugin et al. [66], who reported the case of a 53-year-old man with pneumococcal meningitis who developed numerous ischaemic lesions in the brainstem and basal ganglia, due to parainfectious vasculitis. Clinical and radiological improvement was observed after delayed corticosteroid initiation. Symptomatic vasculitis relapsed after steroid withdrawal and stabilised after reintroduction of the immunosuppressive therapy. While, the CSF contained high levels of MMP-9 at the time of symptomatic vasculitis, a significant decrease of the enzyme accompanied the introduction of corticotherapy and the regression of vasculitic symptoms.

MMP-9 was found to contribute to the pathophysiology of traumatic brain injury. In this regard, MMP-9 knockout mice were shown to have less motor deficits than wild-type mice after controlled cortical impact [72]. At 7 days, traumatic brain lesion volumes on Nissl-stained histological sections were significantly smaller in MMP-9 deficient animals.

MMPs are also associated with neurodegenerative diseases such as Alzheimer's disease and Huntington's disease. The implication of MMPs in neurodegeneration has been further substantiated in a study using an experimental model of striatal degeneration induced by the mitochondrial excitotoxin 3-nitropropionic acid [183]. Using wild type mice, Kim et al. showed an early increase (2 h after toxin injection) of MMP-9 expression in the injured striatum, which was rapidly followed by an increase in BBB permeability. MMP inhibition attenuated BBB disruption, swelling, and lesion volume compared with vehicle-treated controls. One observed a clear spatial relationship between MMP-9 expression and oxidised hydroethidine, indicating reactive oxygen species production. Furthermore, transgenic mice that overexpress SOD1 showed decreased lesion size and oedema along with decreased immunoreactivity for MMP-9, compared with wild-type littermates. On the other hand, knockout mice deficient in SOD1 displayed significantly greater swelling. The authors concluded that early expression and activation of MMP-9 by ROS was involved in early BBB disruption and progressive striatal damage after 3-nitropropionic acid treatment.

Finally, MMPs seem to play a role in some inflammatory myopathies and tumours of the CNS such as glioma [2, 184]. It is likely that the number of neurological diseases associated with unbalanced MMP/TIMP expression will grow over time.

Perspectives

MMPs play a pleiotropic role in the pathophysiology of BBB disruption and delayed neuronal cell death during ischaemic and inflammatory brain injury. Recent studies also suggest that MMPs may play a role in glial scarring, neuronal cell migration and brain tissue recovery. In this context, new therapeutic strategies designed to modulate MMP activity should take into account the multiple levels at which these proteinases act as well as the timing of their involvement in different pathophysiologic processes. An oversimplified experimental approach to new therapeutic interventions using MMP inhibitors would almost certainly bring about disappointing clinical results. In this regard, clinical stroke studies have been most often deceiving. Future studies should dissociate the acute phase of stroke, where MMPs play a deleterious role on the BBB and neuronal cell survival, from the sub-acute phase, when MMPs may play a more beneficial role by favouring neuronal cell migration and recovery.

During the acute phase, therapeutic strategies should aim at either increasing the therapeutic window for thrombolysis or determining biological markers susceptible to help clinicians to evaluate the risk of HT in individual patients. Future clinical studies should evaluate the value of MMPs as marker of the risk of HT and as target to prevent brain haemorrhage.

Our knowledge regarding the role of MMPs in the recovery phase of stroke is currently too weak to build reasonable hypotheses on future therapeutic strategies involving MMPs. Nevertheless, it is likely that upcoming years will see important scientific developments in brain cell transplantation. The ECM is a key modulator of cell migration and survival, future studies interested in cell transplantation will have to unravel ECM biology in order to have a chance to bring positive results. No doubt that MMPs will be shown to play a critical role in the fate of brain cell transplantation.

References

1   Yong VW, Power C, Forsyth P, Edwards DR (2001) Metalloproteinases in biology and pathology of the nervous system. *Nat Rev Neurosci* 2: 502–511
2   Yong VW, Krekoski CA, Forsyth PA, Bell R, Edwards DR (1998) Matrix metalloproteinases and diseases of the CNS. *Trends Neurosci* 21: 75–80
3   Massova I, Kotra LP, Fridman R, Mobashery S (1998) Matrix metalloproteinases: structures, evolution, and diversification. *Faseb J* 12: 1075–1095
4   Sternlicht MD, Werb Z (2001) How matrix metalloproteinases regulate cell behavior. *Annu Rev Cell Dev Biol* 17: 463–516
5   Gasche Y, Fujimura M, Morita-Fujimura Y, Copin JC, Kawase M, Massengale J, Chan PH (1999) Early appearance of activated matrix metalloproteinase-9 after focal cerebral

ischemia in mice: a possible role in blood-brain barrier dysfunction. *J Cereb Blood Flow Metab* 19: 1020–1028

6   Zhang JW, Gottschall PE (1997) Zymographic measurement of gelatinase activity in brain tissue after detergent extraction and affinity-support purification. *J Neurosci Methods* 76: 15–20

7   Fujimura M, Gasche Y, Morita-Fujimura Y, Massengale J, Kawase M, Chan PH (1999) Early appearance of activated matrix metalloproteinase-9 and blood-brain barrier disruption in mice after focal cerebral ischemia and reperfusion. *Brain Res* 842: 92–100

8   Overall CM, Wrana JL, Sodek J (1991) Transcriptional and post-transcriptional regulation of 72-kDa gelatinase/type IV collagenase by transforming growth factor-beta 1 in human fibroblasts. Comparisons with collagenase and tissue inhibitor of matrix metalloproteinase gene expression. *J Biol Chem* 266: 14064–14071

9   Delany AM, Brinckerhoff CE (1992) Post-transcriptional regulation of collagenase and stromelysin gene expression by epidermal growth factor and dexamethasone in cultured human fibroblasts. *J Cell Biochem* 50: 400–410

10   Akool el S, Kleinert H, Hamada FM, Abdelwahab MH, Forstermann U, Pfeilschifter J, Eberhardt W (2003) Nitric oxide increases the decay of matrix metalloproteinase 9 mRNA by inhibiting the expression of mRNA-stabilizing factor HuR. *Mol Cell Biol* 23: 4901–4916

11   Christman JW, Blackwell TS, Juurlink BH (2000) Redox regulation of nuclear factor kappa B: therapeutic potential for attenuating inflammatory responses. *Brain Pathol* 10: 153–162

12   Huhtala P, Chow LT, Tryggvason K (1990) Structure of the human type IV collagenase gene. *J Biol Chem* 265: 11077–11082

13   Brooks PC, Stromblad S, Sanders LC, von Schalscha TL, Aimes RT, Stetler-Stevenson WG, Quigley JP, Cheresh DA (1996) Localization of matrix metalloproteinase MMP-2 to the surface of invasive cells by interaction with integrin alpha v beta 3. *Cell* 85: 683–693

14   Yu Q, Stamenkovic I (2000) Cell surface-localized matrix metalloproteinase-9 proteolytically activates TGF-beta and promotes tumor invasion and angiogenesis. *Genes Dev* 14: 163–176

15   Yu WH, Woessner JF Jr (2000) Heparan sulfate proteoglycans as extracellular docking molecules for matrilysin (matrix metalloproteinase 7). *J Biol Chem* 275: 4183–4191

16   Van Wart HE, Birkedal-Hansen H (1990) The cysteine switch: a principle of regulation of metalloproteinase activity with potential applicability to the entire matrix metalloproteinase gene family. *Proc Natl Acad Sci USA* 87: 5578–5582

17   Stricklin GP, Jeffrey JJ, Roswit WT, Eisen AZ (1983) Human skin fibroblast procollagenase: mechanisms of activation by organomercurials and trypsin. *Biochemistry* 22: 61–68

18   Knauper V, Will H, Lopez-Otin C, Smith B, Atkinson SJ, Stanton H, Hembry RM, Murphy G (1996) Cellular mechanisms for human procollagenase-3 (MMP-13) activation.

Evidence that MT1-MMP (MMP-14) and gelatinase a (MMP-2) are able to generate active enzyme. *J Biol Chem* 271: 17124–17131

19 Rajagopalan S, Meng XP, Ramasamy S, Harrison DG, Galis ZS (1996) Reactive oxygen species produced by macrophage-derived foam cells regulate the activity of vascular matrix metalloproteinases *in vitro*. Implications for atherosclerotic plaque stability. *J Clin Invest* 98: 2572–2579

20 Weiss SJ, Peppin G, Ortiz X, Ragsdale C, Test ST (1985) Oxidative autoactivation of latent collagenase by human neutrophils. *Science* 227: 747–749

21 Tyree B, Seltzer JL, Halme J, Jeffrey JJ, Eisen AZ (1981) The stoichiometric activation of human skin fibroblast pro-collagenase by factors present in human skin and rat uterus. *Arch Biochem Biophys* 208: 440–443

22 Gu Z, Kaul M, Yan B, Kridel SJ, Cui J, Strongin A, Smith JW, Liddington RC, Lipton SA (2002) S-nitrosylation of matrix metalloproteinases: signaling pathway to neuronal cell death. *Science* 297: 1186–1190

23 Kotra LP, Zhang L, Fridman R, Orlando R, Mobashery S (2002) N-Glycosylation pattern of the zymogenic form of human matrix metalloproteinase-9. *Bioorg Chem* 30: 356–370

24 Mignatti P, Rifkin DB (1993) Biology and biochemistry of proteinases in tumor invasion. *Physiol Rev* 73: 161–195

25 Kleiner DE Jr, Stetler-Stevenson WG (1993) Structural biochemistry and activation of matrix metalloproteases. *Curr Opin Cell Biol* 5: 891–897

26 Lohi J, Lehti K, Westermarck J, Kahari VM, Keski-Oja J (1996) Regulation of membrane-type matrix metalloproteinase-1 expression by growth factors and phorbol 12-myristate 13-acetate. *Eur J Biochem* 239: 239–247

27 Bergmann U, Tuuttila A, Stetler-Stevenson WG, Tryggvason K (1995) Autolytic activation of recombinant human 72 kilodalton type IV collagenase. *Biochemistry* 34: 2819–2825

28 Goldberg GI, Marmer BL, Grant JA, Eisen AZ, Wilhelm S, He C (1989) Human 72k type IV collagenase forms a complex with a tissue inhibitor of metalloproteinase designed TIMP-2. *Proc Natl Acad Sci USA* 86: 8207–8211

29 Strongin AY, Collier Y, Bannikov G, Marmer BL, Grant GA, Goldberg GI (1995) Mechanism of cell surface activation of 72-kDa type IV collagenase. *J Biol Chem* 270: 5331–5338

30 Strongin AY, Marmer BL, Grant GA, Goldberg GI (1993) Plasma membrane-dependent activation of the 72-kDa type IV collagenase is prevented by complex formation with TIMP-2. *J Biol Chem* 268: 14033–14039

31 Willenbrock F, Murphy G (1994) Structure-function relationships in the tissue inhibitors of metalloproteinases. *Am J Respir Crit Care Med* 150: S165–S170

32 Dzwonek J, Rylski M, Kaczmarek L (2004) Matrix metalloproteinases and their endogenous inhibitors in neuronal physiology of the adult brain. *FEBS Lett* 567: 129–135

33 Vaillant C, Didier-Bazes M, Hutter A, Belin MF, Thomasset N (1999) Spatiotemporal

expression patterns of metalloproteinases and their inhibitors in the postnatal developing rat cerebellum. *J Neurosci* 19: 4994–5004

34 Wright JW, Masino AJ, Reichert JR, Turner GD, Meighan SE, Meighan PC, Harding JW (2003) Ethanol-induced impairment of spatial memory and brain matrix metalloproteinases. *Brain Res* 963: 252–261

35 Zhang JW, Deb S, Gottschall PE (1998) Regional and differential expression of gelatinases in rat brain after systemic kainic acid or bicuculline administration. *Eur J Neurosci* 10: 3358–3368

36 Zhang JW, Deb S, Gottschall PE (2000) Regional and age-related expression of gelatinases in the brains of young and old rats after treatment with kainic acid. *Neurosci Lett* 295: 9–12

37 Romanic AM, White RF, Arleth AJ, Ohlstein EH, Barone FC (1998) Matrix metalloproteinase expression increases after cerebral focal ischemia in rats: inhibition of matrix metalloproteinase-9 reduces infarct size. *Stroke* 29: 1020–1030

38 Rosenberg GA, Navratil M, Barone F, Feuerstein G (1996) Proteolytic cascade enzymes increase in focal cerebral ischemia in rat. *J Cereb Blood Flow Metab* 16: 360–366

39 Asahi M, Asahi K, Jung JC, del Zoppo GJ, Fini ME, Lo EH (2000) Role for matrix metalloproteinase 9 after focal cerebral ischemia: effects of gene knockout and enzyme inhibition with BB-94. *J Cereb Blood Flow Metab* 20: 1681–1689

40 Heo JH, Lucero J, Abumiya T, Koziol JA, Copeland BR, del Zoppo GJ (1999) Matrix metalloproteinases increase very early during experimental focal cerebral ischemia. *J Cereb Blood Flow Metab* 19: 624–633

41 Anthony DC, Ferguson B, Matyzak MK, Miller KM, Esiri MM, Perry VH (1997) Differential matrix metalloproteinase expression in cases of multiple sclerosis and stroke. *Neuropathol Appl Neurobiol* 23: 406–415

42 Clark AW, Krekoski CA, Bou SS, Chapman KR, Edwards DR (1997) Increased gelatinase A (MMP-2) and gelatinase B (MMP-9) activities in human brain after focal ischemia. *Neurosci Lett* 238: 53–56

43 Lewen A, Matz P, Chan PH (2000) Free radical pathways in CNS injury. *J Neurotrauma* 17: 871–890

44 Gasche Y, Copin JC, Sugawara T, Fujimura M, Chan PH (2001) Matrix metalloproteinase inhibition prevents oxidative stress-associated blood-brain barrier disruption after transient focal cerebral ischemia. *J Cereb Blood Flow Metab* 21: 1393–1400

45 Morita-Fujimura Y, Fujimura M, Gasche Y, Copin JC, Chan PH (2000) Overexpression of copper and zinc superoxide dismutase in transgenic mice prevents the induction and activation of matrix metalloproteinases after cold injury-induced brain trauma. *J Cereb Blood Flow Metab* 20: 130–138

46 Gidday JM, Gasche YG, Copin JC, Shah AR, Perez RS, Shapiro SD, Chan PH, Park TS (2005) Leukocyte-derived matrix metalloproteinase-9 mediates blood-brain barrier breakdown and is proinflammatory following transient focal cerebral ischemia. *Am J Physiol Heart Circ Physiol* 289: H558–568

47 Rosenberg GA, Cunningham LA, Wallace J, Alexander S, Estrada EY, Grossetete M,

Razhagi A, Miller K, Gearing A (2001) Immunohistochemistry of matrix metallopro-teinases in reperfusion injury to rat brain: activation of MMP-9 linked to stromelysin-1 and microglia in cell cultures. *Brain Res* 893: 104–112

48   Rosenberg GA, Estrada EY, Dencoff JE (1998) Matrix metalloproteinases and TIMPs are associated with blood-brain barrier opening after reperfusion in rat brain. *Stroke* 29: 2189–2195

49   Wang X, Barone FC, White RF, Feuerstein GZ (1998) Subtractive cloning identifies tissue inhibitor of matrix metalloproteinase-1 (TIMP-1) increased gene expression fol-lowing focal stroke. *Stroke* 29: 516–520

50   Nguyen M, Arkwell J, Jackson CJ (1998) Active and tissue inhibitor of matrix metal-loproteinase-free gelatinase B accumulates within human microvascular endothelial vesicles. *J Biol Chem* 273: 5400–5404

51   Cossins JA, Clements JM, Ford J, Miller KM, Pigott R, Vos W, Van der Valk P, De Groot CJ (1997) Enhanced expression of MMP-7 and MMP-9 in demyelinating multiple scle-rosis lesions. *Acta Neuropathol (Berl)* 94: 590–598

52   Maeda A, Sobel RA (1996) Matrix metalloproteinases in the normal human central nervous system, microglial nodules, and multiple sclerosis lesions. *J Neuropathol Exp Neurol* 55: 300–309

53   Kieseier BC, Kiefer R, Clements JM, Miller K, Wells GM, Schweitzer T, Gearing AJ, Hartung HP (1998) Matrix metalloproteinase-9 and -7 are regulated in experimental autoimmune encephalomyelitis. *Brain* 121(Pt 1): 159–166

54   Gijbels K, Masure S, Carton H, Opdenakker G (1992) Gelatinase in the cerebrospinal fluid of patients with multiple sclerosis and other inflammatory neurological disorders. *J Neuroimmunol* 41: 29–34

55   Leppert D, Ford J, Stabler G, Grygar C, Lienert C, Huber S, Miller KM, Hauser SL, Kappos L (1998) Matrix metalloproteinase-9 (gelatinase B) is selectively elevated in CSF during relapses and stable phases of multiple sclerosis. *Brain* 121 (Pt 12): 2327–2334

56   Lee MA, Palace J, Stabler G, Ford J, Gearing A, Miller K (1999) Serum gelatinase B, TIMP-1 and TIMP-2 levels in multiple sclerosis. A longitudinal clinical and MRI study. *Brain* 122 (Pt 2): 191–197

57   Waubant E, Goodkin DE, Gee L, Bacchetti P, Sloan R, Stewart T, Andersson PB, Stabler G, Miller K (1999) Serum MMP-9 and TIMP-1 levels are related to MRI activity in relapsing multiple sclerosis. *Neurology* 53: 1397–1401

58   Trojano M, Avolio C, Liuzzi GM, Ruggieri M, Defazio G, Liguori M, Santacroce MP, Paolicelli D, Giuliani F, Riccio P et al (1999) Changes of serum sICAM-1 and MMP-9 induced by rIFNbeta-1b treatment in relapsing-remitting MS. *Neurology* 53: 1402–1408

59   Lichtinghagen R, Seifert T, Kracke A, Marckmann S, Wurster U, Heidenreich F (1999) Expression of matrix metalloproteinase-9 and its inhibitors in mononuclear blood cells of patients with multiple sclerosis. *J Neuroimmunol* 99: 19–26

60   Cuzner ML, Gveric D, Strand C, Loughlin AJ, Paemen L, Opdenakker G, Newcombe J (1996) The expression of tissue-type plasminogen activator, matrix metalloproteases

and endogenous inhibitors in the central nervous system in multiple sclerosis: comparison of stages in lesion evolution. *J Neuropathol Exp Neurol* 55: 1194–1204

61 Galboiz Y, Shapiro S, Lahat N, Rawashdeh H, Miller A (2001) Matrix metalloproteinases and their tissue inhibitors as markers of disease subtype and response to interferon-beta therapy in relapsing and secondary-progressive multiple sclerosis patients. *Ann Neurol* 50: 443–451

62 Kouwenhoven M, Ozenci V, Tjernlund A, Pashenkov M, Homman M, Press R, Link H (2002) Monocyte-derived dendritic cells express and secrete matrix-degrading metalloproteinases and their inhibitors and are imbalanced in multiple sclerosis. *J Neuroimmunol* 126: 161–171

63 Ozenci V, Rinaldi L, Teleshova N, Matusevicius D, Kivisakk P, Kouwenhoven M, Link H (1999) Metalloproteinases and their tissue inhibitors in multiple sclerosis. *J Autoimmun* 12: 297–303

64 Paul R, Angele B, Sporer B, Pfister HW, Koedel U (2004) Inflammatory response during bacterial meningitis is unchanged in Fas- and Fas ligand-deficient mice. *J Neuroimmunol* 152: 78–82

65 Paul R, Lorenzl S, Koedel U, Sporer B, Vogel U, Frosch M, Pfister HW (1998) Matrix metalloproteinases contribute to the blood-brain barrier disruption during bacterial meningitis. *Ann Neurol* 44: 592–600

66 Pugin D, Copin J-C, Goodyear M-C, Landis T, Gasche Y (2006) Persisting vasculitis after pneumococcal meningitis: a possible role for matrix metalloproteinase-9 as a marker of disease. *Neurocritical Care* 4: 237–240

67 Leppert D, Leib SL, Grygar C, Miller KM, Schaad UB, Hollander GA (2000) Matrix metalloproteinase (MMP)-8 and MMP-9 in cerebrospinal fluid during bacterial meningitis: association with blood-brain barrier damage and neurological sequelae. *Clin Infect Dis* 31: 80–84

68 Leib SL, Clements JM, Lindberg RL, Heimgartner C, Loeffler JM, Pfister LA, Tauber MG, Leppert D (2001) Inhibition of matrix metalloproteinases and tumour necrosis factor alpha converting enzyme as adjuvant therapy in pneumococcal meningitis. *Brain* 124: 1734–1742

69 Leib SL, Leppert D, Clements J, Tauber MG (2000) Matrix metalloproteinases contribute to brain damage in experimental pneumococcal meningitis. *Infect Immun* 68: 615–620

70 Khuth ST, Akaoka H, Pagenstecher A, Verlaeten O, Belin MF, Giraudon P, Bernard A (2001) Morbillivirus infection of the mouse central nervous system induces region-specific upregulation of MMPs and TIMPs correlated to inflammatory cytokine expression. *J Virol* 75: 8268–8282

71 Szklarczyk A, Lapinska J, Rylski M, McKay RD, Kaczmarek L (2002) Matrix metalloproteinase-9 undergoes expression and activation during dendritic remodeling in adult hippocampus. *J Neurosci* 22: 920–930

72 Wang X, Jung J, Asahi M, Chwang W, Russo L, Moskowitz MA, Dixon CE, Fini ME,

Lo EH (2000) Effects of matrix metalloproteinase-9 gene knock-out on morphological and motor outcomes after traumatic brain injury. *J Neurosci* 20: 7037–7042

73 Zagulska-Szymczak S, Filipkowski RK, Kaczmarek L (2001) Kainate-induced genes in the hippocampus: lessons from expression patterns. *Neurochem Int* 38: 485–501

74 Jaworowicz DJ, Korytko PJ, Singh Lakhman S, Boje KM (1998) Nitric oxide and prostaglandin E2 formation parallels blood-brain barrier disruption in an experimental rat model of bacterial meningitis. *Brain Res Bull* 46: 541–546

75 Nedivi E, Hevroni D, Naot D, Israeli D, Citri Y (1993) Numerous candidate plasticity-related genes revealed by differential cDNA cloning. *Nature* 363: 718–722

76 Schönbeck U, Mach F, Liby P (1998) Generation of biologically active IL-1β by matrix metalloproteinases: a novel caspase-1-independent pathway of IL-1β processing. *J Immunol* 161: 3340–3346

77 Schwab S, Steiner T, Aschoff A, Schwarz S, Steiner HH, Jansen O, Hacke W (1998) Early hemicraniectomy in patients with complete middle cerebral artery infarction. *Stroke* 29: 1888–1893

78 Clark WM, Albers GW, Madden KP, Hamilton S (2000) The rtPA (alteplase) 0- to 6-hour acute stroke trial, part A (A0276g): results of a double-blind, placebo-controlled, multicenter study. Thromblytic therapy in acute ischemic stroke study investigators. *Stroke* 31: 811–816

79 Latour LL, Kang DW, Ezzeddine MA, Chalela JA, Warach S (2004) Early blood-brain barrier disruption in human focal brain ischemia. *Ann Neurol* 56: 468–477

80 Neumann-Haefelin T, Kastrup A, de Crespigny A, Yenari MA, Ringer T, Sun GH, Moseley ME (2000) Serial MRI after transient focal cerebral ischemia in rats dynamics of tissue injury, blood-brain barrier damage, and edema formation. *Stroke* 31: 1965–1973

81 Copin JC, Goodyear MC, Gidday JM, Shah AR, Morel DR, Gasche Y (2005) Role of matrix metalloproteinases in apoptosis following focal cerebral ischemia in rat and mice. *Eur J Neurosci* 22: 1597–1608

82 Barone FC, Feuerstein GZ (1999) Inflammatory mediators and stroke: new opportunities for novel therapeutics. *J Cereb Blood Flow Metab* 19: 819–834

83 Petty MA, Wettstein JG (2001) Elements of cerebral microvascular ischaemia. *Brain Res Brain Res Rev* 36: 23–34

84 Clark WM, Lutsep HL (2001) Potential of anticytokine therapies in central nervous system ischaemia. *Expert Opin Biol Ther* 1: 227–237

85 Hosomi N, Ban CR, Naya T, Takahashi T, Guo P, Song XY, Kohno M (2005) Tumor necrosis factor-alpha neutralization reduced cerebral edema through inhibition of matrix metalloproteinase production after transient focal cerebral ischemia. *J Cereb Blood Flow Metab* 25: 959–967

86 del Zoppo GJ (1994) Microvascular changes during cerebral ischemia and reperfusion. *Cerebrovasc Brain Metab Rev* 6: 47–96

87 Okada Y, Copeland BR, Mori E (1994) P-selectin and intercellular adhesion molecule-1 expression after focal brain ischemia and reperfusion. *Stroke* 25: 202–211

88   Yurchenco PD, Schittny JC (1990) Molecular architecture of basement membranes. *FASEB J* 4: 1577–1590

89   Brooks PC, Montgomery AM, Rosenfeld M, Reisfeld RA, Hu T, Klier G, Cheresh DA (1994) Integrin alpha v beta 3 antagonists promote tumor regression by inducing apoptosis of angiogenic blood vessels. *Cell* 79: 1157–1164

90   Chen CS, Mrksich M, Huang S, Whitesides GM, Ingber DE (1997) Geometric control of cell life and death. *Science* 276: 1425–1428

91   Ingber DE, Folkman J (1989) How does extracellular matrix control capillary morphogenesis. *Cell* 58: 803–805

92   Maniotis AJ, Chen CS, Ingber DE (1997) Demonstration of mechanical connections between integrins cytoskeletal filaments and nucleoplasm that stabilize nuclear structure. *Proc Natl Acad Sci* 94: 849–854

93   Hamann GF, Okada Y, Fitridge R, Del Zoppo GJ (1995) Microvascular basal lamina antigens disappear during cerebral ischemia and reperfusion. *Stroke* 26: 2120–2126

94   Belayev L, Busto R, Zhao W, Ginsberg MD (1996) Quantitative evaluation of blood-brain barrier permeability following middle cerebral artery occlusion in rats. *Brain Res* 739: 88–96

95   Kondo T, Reaume AG, Huang TT, Carlson E, Murakami K, Chen SF, Hoffman EK, Scott RW, Epstein CJ, Chan PH (1997) Reduction of CuZn-superoxide dismutase activity exacerbates neuronal cell injury and edema formation after transient focal cerebral ischemia. *J Neurosci* 17: 4180–4189

96   Hamann GF, del Zoppo GJ, von Kummer R (1999) Hemorrhagic transformation of cerebral infarction–possible mechanisms. *Thromb Haemost* 82: 92–94

97   Hamann GF, Okada Y, Del Zoppo GJ (1996) Hemorrhagic transformation and microvascular integrity during focal cerebral ischemi/reperfusion. *J Cereb Blood Flow Metab* 16: 1373–1378

98   Asahi M, Wang X, Mori T, Sumii T, Jung JC, Moskowitz MA, Fini ME, Lo EH (2001) Effects of matrix metalloproteinase-9 gene knock-out on the proteolysis of blood-brain barrier and white matter components after cerebral ischemia. *J Neurosci* 21: 7724–7732

99   Asahi M, Sumii T, Fini ME, Itohara S, Lo EH (2001) Matrix metalloproteinase 2 gene knockout has no effect on acute brain injury after focal ischemia. *Neuroreport* 12: 3003–3007

100 Fukuda S, Fini CA, Mabuchi T, Koziol JA, Eggleston LL Jr, del Zoppo GJ (2004) Focal cerebral ischemia induces active proteases that degrade microvascular matrix. *Stroke* 35: 998–1004

101 Yang YI, Estrada EY, Thompson JF, Liu W, Rosenberg GA (2007) Matrix metalloproteinase-mediated disruption of tight junction proteins in cerebral vessels is reversed by synthetic matrix metalloproteinase inhibitor in focal ischemia in rat. *J Cereb Blood Flow Metab* 27: 697–709

102 Opdenakker G, Van den Steen PE, Dubois B, Nelissen I, Van Coillie E, Masure S, Proost

P, Van Damme J (2001) Gelatinase B functions as regulator and effector in leukocyte biology. *J Leukoc Biol* 69: 851–859

103 Asahi M, Asahi K, Wang X, Lo EH (2000) Reduction of tissue plasminogen activator-induced hemorrhage and brain injury by free radical spin trapping after embolic focal cerebral ischemia in rats. *J Cereb Blood Flow Metab* 20: 452–457

104 Gursoy-Ozdemir Y, Can A, Dalkara T (2004) Reperfusion-induced oxidative/nitrative injury to neurovascular unit after focal cerebral ischemia. *Stroke* 35: 1449–1453

105 Lewen A, Sugawara T, Gasche Y, Fujimura M, Chan PH (2001) Oxidative cellular damage and the reduction of APE/Ref-1 expression after experimental traumatic brain injury. *Neurobiol Dis* 8: 380–390

106 Maier CM, Hsieh L, Crandall T, Narasimhan P, Chan PH (2006) Evaluating therapeutic targets for reperfusion-related brain hemorrhage. *Ann Neurol* 59: 929–938

107 Lapchak PA, Chapman DF, Zivin JA (2000) Metalloproteinase inhibition reduces thrombolytic (tissue plasminogen activator)-induced hemorrhage after thromboembolic stroke. *Stroke* 31: 3034–3040

108 Pfefferkorn T, Rosenberg GA (2003) Closure of the blood-brain barrier by matrix metalloproteinase inhibition reduces rtPA-mediated mortality in cerebral ischemia with delayed reperfusion. *Stroke* 34: 2025–2030

109 Zhang RL, Chopp M, Zhang ZG, Jiang Q, Ewing JR (1997) A rat model of focal embolic cerebral ischemia. *Brain Res* 766: 83–92

110 Wang CX, Yang T, Shuaib A (2001) An improved version of embolic model of brain ischemic injury in the rat. *J Neurosci Methods* 109: 147–151

111 Sumii T, Lo EH (2002) Involvement of matrix metalloproteinase in thrombolysis-associated hemorrhagic transformation after embolic focal ischemia in rats. *Stroke* 33: 831–836

112 Wang X, Lee SR, Arai K, Lee SR, Tsuji K, Rebeck GW, Lo EH (2003) Lipoprotein receptor-mediated induction of matrix metalloproteinase by tissue plasminogen activator. *Nat Med* 9: 1313–1317

113 Yepes M, Sandkvist M, Moore EG, Bugge TH, Strickland DK, Lawrence DA (2003) Tissue-type plasminogen activator induces opening of the blood-brain barrier *via* the LDL receptor-related protein. *J Clin Invest* 112: 1533–1540

114 Wang J, Tsirka SE (2005) Neuroprotection by inhibition of matrix metalloproteinases in a mouse model of intracerebral haemorrhage. *Brain* 128: 1622–1633

115 Svedin P, Hagberg H, Sävman K, Zhu C, Mallard C (2007) Matrix metalloproteinase-9 gene knock-out protects the immature brain after cerebral hypoxia-ischemia. *J Neurosci* 27: 1511–1518

116 Montaner J, Molina CA, Monasterio J, Abilleira S, Arenillas JF, Ribo M, Quintana M, Alvarez-Sabin J (2003) Matrix metalloproteinase-9 pretreatment level predicts intracranial hemorrhagic complications after thrombolysis in human stroke. *Circulation* 107: 598–603

117 Castellanos M, Sobrino T, Millan M, Garcia M, Arenillas J, Nombela F, Brea D, Perez de la Ossa N, Serena J, Vivancos J et al (2007) Serum cellular fibronectin and matrix

metalloproteinase-9 as screening biomarkers for the prediction of parenchymal hematoma after thrombolytic therapy in acute ischemic stroke. A multicenter confirmatory study. *Stroke* 38: 1855–1859

118 Montaner J, Alvarez-Sabin J, Molina C, Angles A, Abilleira S, Arenillas J, Gonzalez MA, Monasterio J (2001) Matrix metalloproteinase expression after human cardioembolic stroke: temporal profile and relation to neurological impairment. *Stroke* 32: 1759–1766

119 Montaner J, Alvarez-Sabin J, Molina CA, Angles A, Abilleira S, Arenillas J, Monasterio J (2001) Matrix metalloproteinase expression is related to hemorrhagic transformation after cardioembolic stroke. *Stroke* 32: 2762–2767

120 Backstrom JR, Lim GP, Cullen MJ, Tokes ZA (1996) Matrix metalloproteinase-9 (MMP-9) is synthesized in neurons of the human hippocampus and is capable of degrading the amyloid-beta peptide (1–40). *J Neurosci* 16: 7910–7919

121 Massengale JL, Gasche Y, Chan PH (2002) Carbohydrate source influences gelatinase production by mouse astrocytes *in vitro*. *Glia* 38: 240–245

122 Gottschall PE, Yu X, Bing B (1995) Increased production of gelatinase B (matrix metalloproteinase-9) and interleukin-6 by activated rat microglia in culture. *J Neurosci Res* 42: 335–342

123 Oh LY, Larsen PH, Krekoski CA, Edwards DR, Donovan F, Werb Z, Yong VW (1999) Matrix metalloproteinase-9/gelatinase B is required for process outgrowth by oligodendrocytes. *J Neurosci* 19: 8464–8475

124 Investigators EAST (2001) Use of anti-ICAM-1 therapy in ischemic stroke: results of the Enlimomab Acute Stroke Trial. *Neurology* 57: 1428–1434

125 Furuya K, Takeda H, Azhar S, McCarron RM, Chen Y, Ruetzler CA, Wolcott KM, DeGraba TJ, Rothlein R, Hugli TE et al (2001) Examination of several potential mechanisms for the negative outcome in a clinical stroke trial of enlimomab, a murine anti-human intercellular adhesion molecule-1 antibody: a bedside-to-bench study. *Stroke* 32: 2665–2674

126 Gursoy-Ozdemir Y, Qiu J, Matsuoka N, Bolay H, Bermpohl D, Jin H, Wang X, Rosenberg GA, Lo EH, Moskowitz MA (2004) Cortical spreading depression activates and upregulates MMP-9. *J Clin Invest* 113: 1447–1455

127 Ferri KF, Kroemer G (2001) Organelle-specific initiation of cell death pathways. *Nat Cell Biol* 3: E255–263

128 Ferri KF, Kroemer G (2001) Mitochondria–the suicide organelles. *Bioessays* 23: 111–115

129 Wang H, Yu SW, Koh DW, Lew J, Coombs C, Bowers W, Federoff HJ, Poirier GG, Dawson TM, Dawson VL (2004) Apoptosis-inducing factor substitutes for caspase executioners in NMDA-triggered excitotoxic neuronal death. *J Neurosci* 24: 10963–10973

130 Yu SW, Wang H, Dawson TM, Dawson VL (2003) Poly(ADP-ribose) polymerase-1 and apoptosis inducing factor in neurotoxicity. *Neurobiol Dis* 14: 303–317

131 Plesnila N, Zinkel S, Le DA, Amin-Hanjani S, Wu Y, Qiu J, Chiarugi A, Thomas SS, Kohane DS, Korsmeyer SJ et al (2001) BID mediates neuronal cell death after oxygen/

glucose deprivation and focal cerebral ischemia. *Proc Natl Acad Sci USA* 98: 15318–15323

132 Wallach D, Varfolomeev EE, Malinin NL, Goltsev YV, Kovalenko AV, Boldin MP (1999) Tumor necrosis factor receptor and Fas signaling mechanisms. *Annu Rev Immunol* 17: 331–367

133 Vogelstein B, Lane D, Levine AJ (2000) Surfing the p53 network. *Nature* 408: 307–310

134 Lipton P (1999) Ischemic cell death in brain neurons. *Physiol Rev* 79: 1431–1568

135 Hara H, Friedlander RM, Gagliardini V, Ayata C, Fink K, Huang Z, Shimizu-Sasamata M, Yuan J, Moskowitz MA (1997) Inhibition of interleukin 1beta converting enzyme family proteases reduces ischemic and excitotoxic neuronal damage. *Proc Natl Acad Sci USA* 94: 2007–2012

136 Sugawara T, Noshita N, Lewen A, Gasche Y, Ferrand-Drake M, Fujimura M, Morita-Fujimura Y, Chan PH (2002) Overexpression of copper/zinc superoxide dismutase in transgenic rats protects vulnerable neurons against ischemic damage by blocking the mitochondrial pathway of caspase activation. *J Neurosci* 22: 209–217

137 Jourquin J, Tremblay E, Decanis N, Charton G, Hanessian S, Chollet AM, Le Diguardher T, Khrestchatisky M, Rivera S (2003) Neuronal activity-dependent increase of net matrix metalloproteinase activity is associated with MMP-9 neurotoxicity after kainate. *Eur J Neurosci* 18: 1507–1517

138 Lee SR, Tsuji K, Lee SR, Lo EH (2004) Role of matrix metalloproteinases in delayed neuronal damage after transient global cerebral ischemia. *J Neurosci* 24: 671–678

139 Tan HK, Heywood D, Ralph GS, Bienemann A, Baker AH, Uney JB (2003) Tissue inhibitor of metalloproteinase 1 inhibits excitotoxic cell death in neurons. *Mol Cell Neurosci* 22: 98–106

140 Wallace JA, Alexander S, Estrada EY, Hines C, Cunningham LA, Rosenberg GA (2002) Tissue inhibitor of metalloproteinase-3 is associated with neuronal death in reperfusion injury. *J Cereb Blood Flow Metab* 22: 1303–1310

141 Wetzel M, Rosenberg GA, Cunningham LA (2003) Tissue inhibitor of metalloproteinases-3 and matrix metalloproteinase-3 regulate neuronal sensitivity to doxorubicin-induced apoptosis. *Eur J Neurosci* 18: 1050–1060

142 Fawcett JW, Asher RA (1999) The glial scar and central nervous system repair. *Brain Res Bull* 49: 377–391

143 Ridet JL, Malhotra SK, Privat A, Gage FH (1997) Reactive astrocytes: cellular and molecular cues to biological function. *Trends Neurosci* 20: 570–577

144 Levine JM, Reynolds R, Fawcett JW (2001) The oligodendrocyte precursor cell in health and disease. *Trends Neurosci* 24: 39–47

145 Kreutzberg GW (1996) Microglia: a sensor for pathological events in the CNS. *Trends Neurosci* 19: 312–318

146 McGraw J, Hiebert GW, Steeves JD (2001) Modulating astrogliosis after neurotrauma. *J Neurosci Res* 63: 109–115

147 Campbell IL, Pagenstecher A (1999) Matrix metalloproteinases and their inhibitors in

the nervous system: the good, the bad and the enigmatic. *Trends Neurosci* 22: 285–287

148 Murphy G, Gavrilovic J (1999) Proteolysis and cell migration: creating a path? *Curr Opin Cell Biol* 11: 614–621

149 Ferguson TA, Muir D (2000) MMP-2 and MMP-9 increase the neurite-promoting potential of schwann cell basal laminae and are upregulated in degenerated nerve. *Mol Cell Neurosci* 16: 157–167

150 Duchossoy Y, Horvat JC, Stettler O (2001) MMP-related gelatinase activity is strongly induced in scar tissue of injured adult spinal cord and forms pathways for ingrowing neurites. *Mol Cell Neurosci* 17: 945–956

151 Ellison JA, Velier JJ, Spera P, Jonak ZL, Wang X, Barone FC, Feuerstein GZ (1998) Osteopontin and its integrin receptor alpha(v)beta3 are upregulated during formation of the glial scar after focal stroke. *Stroke* 29: 1698–1706; discussion 1707

152 Agnihotri R, Crawford HC, Haro H, Matrisian LM, Havrda MC, Liaw L (2001) Osteopontin, a novel substrate for matrix metalloproteinase-3 (stromelysin-1) and matrix metalloproteinase-7 (matrilysin). *J Biol Chem* 276: 28261–28267

153 Copin JC, Gasche Y (2007) Matrix metalloproteinase-9 deficiency has no effect on glial scar formation after transient focal cerebral ischemia in mouse. *Brain Res* 1150: 167–173

154 Zhao BQ, Wang S, Kim HY, Storrie H, Rosen BR, Mooney DJ, Wang X, Lo EH (2006) Role of matrix metalloproteinases in delayed cortical responses after stroke. *Nat Med* 12: 441–445

155 Sood RR, Taheri S, Candelario-Jalil E, Estrada EY, Rosenberg GA (2008) Early beneficial effect of matrix metalloproteinase inhibition on blood–brain barrier permeability as measured by magnetic resonance imaging countered by impaired long-term recovery after stroke in rat brain. *J Cereb Blood Flow Metab* 28: 431–438

156 Lampl Y, Boaz M, Gilad R, Lorberboym M, Dabby R, Rapoport A, Anca-Hershkowitz M, Sadeh M (2007) Minocycline treatment in acute stroke: An open-label, evaluator-blinded study. *Neurology* 69: 1404–1410

157 Koistinaho M, Malm TM, Kettunen MI, Goldsteins G, Starckx S, Kauppinen RA, Opdenakker G, Koistinaho J (2005) Minocycline protects against permanent cerebral ischemia in wild type but not in matrix metalloprotease-9-deficient mice. *J Cereb Blood Flow Metab* 25: 460–467

158 Jordan J, Fernandez-Gomez FJ, Ramos M, Ikuta I, Aguirre N, Galindo MF (2007) Minocycline and cytoprotection: shedding new light on a shadowy controversy. *Current Drug Delivery* 4: 225–231

159 Wasserman JK, Schlichter LC (2007) Minocycline protects the blood–brain barrier and reduces intracerebral hemorrhage in the rat. *Experimental Neurol* 207: 227–237

160 Matyszak MK, Perry VH (1996) Delayed-type hypersensitivity lesions in the central nervous system are prevented by inhibitors of matrix metalloproteinases. *J Neuroimmunol* 69: 141–149

161 Chandler S, Miller KM, Clements JM, Lury J, Corkill D, Anthony DC, Adams SE, Gear-

ing AJ (1997) Matrix metalloproteinases, tumor necrosis factor and multiple sclerosis: an overview. *J Neuroimmunol* 72: 155–161

162 Opdenakker G, Van Damme J (1994) Cytokine-regulated proteases in autoimmune diseases. *Immunol Today* 15: 103–107

163 Dubois B, Masure S, Hurtenbach U, Paemen L, Heremans H, van den Oord J, Sciot R, Meinhardt T, Hammerling G, Opdenakker G et al (1999) Resistance of young gelatinase B-deficient mice to experimental autoimmune encephalomyelitis and necrotizing tail lesions. *J Clin Invest* 104: 1507–1515

164 Lou J, Gasche Y, Zheng L, Giroud C, Morel P, Clements J, Ythier A, Grau GE (1999) Interferon-beta inhibits activated leukocyte migration through human brain microvascular endothelial cell monolayer. *Lab Invest* 79: 1015–1025

165 Leppert D, Waubant E, Burk MR, Oksenberg JR, Hauser SL (1996) Interferon beta-1b inhibits gelatinase secretion and *in vitro* migration of human T cells: a possible mechanism for treatment efficacy in multiple sclerosis. *Ann Neurol* 40: 846–852

166 Stuve O, Dooley NP, Uhm JH, Antel JP, Francis GS, Williams G, Yong VW (1996) Interferon beta-1b decreases the migration of T lymphocytes *in vitro*: effects on matrix metalloproteinase-9. *Ann Neurol* 40: 853–863

167 Ozenci V, Kouwenhoven M, Teleshova N, Pashenkov M, Fredrikson S, Link H (2000) Multiple sclerosis: pro- and anti-inflammatory cytokines and metalloproteinases are affected differentially by treatment with IFN-beta. *J Neuroimmunol* 108: 236–243

168 Nelissen I, Martens E, Van den Steen PE, Proost P, Ronsse I, Opdenakker G (2003) Gelatinase B/matrix metalloproteinase-9 cleaves interferon-beta and is a target for immunotherapy. *Brain* 126: 1371–1381

169 Shipley JM, Wesselschmidt RL, Kobayashi DK, Ley TJ, Shapiro SD (1996) Metalloelastase is required for macrophage-mediated proteolysis and matrix invasion in mice. *Proc Natl Acad Sci USA* 93: 3942–3946

170 Durand ML, Calderwood SB, Weber DJ, Miller SI, Southwick FS, Caviness VS Jr, Swartz MN (1993) Acute bacterial meningitis in adults. A review of 493 episodes. *N Engl J Med* 328: 21–28

171 Kastenbauer S, Pfister HW (2003) Pneumococcal meningitis in adults: spectrum of complications and prognostic factors in a series of 87 cases. *Brain* 126: 1015–1025

172 Pfister HW, Borasio GD, Dirnagl U, Bauer M, Einhaupl KM (1992) Cerebrovascular complications of bacterial meningitis in adults. *Neurology* 42: 1497–1504

173 Pfister HW, Feiden W, Einhaupl KM (1993) Spectrum of complications during bacterial meningitis in adults. Results of a prospective clinical study. *Arch Neurol* 50: 575–581

174 van de Beek D, de Gans J, McIntyre P, Prasad K (2004) Steroids in adults with acute bacterial meningitis: a systematic review. *Lancet Infect Dis* 4: 139–143

175 Tauber MG, Moser B (1999) Cytokines and chemokines in meningeal inflammation: biology and clinical implications. *Clin Infect Dis* 28: 1–11; quiz 12

176 Leib SL, Tauber MG (1999) Pathogenesis of bacterial meningitis. *Infect Dis Clin North Am* 13: 527–548, v–vi

177 Meli DN, Christen S, Leib SL (2003) Matrix metalloproteinase-9 in pneumococcal meningitis: activation *via* an oxidative pathway. *J Infect Dis* 187: 1411–1415

178 Meli DN, Loeffler JM, Baumann P, Neumann U, Buhl T, Leppert D, Leib SL (2004) In pneumococcal meningitis a novel water-soluble inhibitor of matrix metalloproteinases and TNF-alpha converting enzyme attenuates seizures and injury of the cerebral cortex. *J Neuroimmunol* 151: 6–11

179 Nau R, Bruck W (2002) Neuronal injury in bacterial meningitis: mechanisms and implications for therapy. *Trends Neurosci* 25: 38–45

180 Nau R, Wellmer A, Soto A, Koch K, Schneider O, Schmidt H, Gerber J, Michel U, Bruck W (1999) Rifampin reduces early mortality in experimental *Streptococcus pneumoniae* meningitis. *J Infect Dis* 179: 1557–1560

181 Williams PL, Leib SL, Kamberi P, Leppert D, Sobel RA, Bifrare YD, Clemons KV, Stevens DA (2002) Levels of matrix metalloproteinase-9 within cerebrospinal fluid in a rabbit model of coccidioidal meningitis and vasculitis. *J Infect Dis* 186: 1692–1695

182 Bottcher T, Spreer A, Azeh I, Nau R, Gerber J (2003) Matrix metalloproteinase-9 deficiency impairs host defense mechanisms against *Streptococcus pneumoniae* in a mouse model of bacterial meningitis. *Neurosci Lett* 338: 201–204

183 Kim GW, Gasche Y, Grzeschik S, Copin JC, Maier CM, Chan PH (2003) Neurodegeneration in striatum induced by the mitochondrial toxin 3-nitropropionic acid: role of matrix metalloproteinase-9 in early blood-brain barrier disruption? *J Neurosci* 23: 8733–8742

184 Nakada M, Okada Y, Yamashita J (2003) The role of matrix metalloproteinases in glioma invasion. *Front Biosci* 8: e261–269

Extracellular matrix remodelling and matrix metalloproteinases in the liver

Bruno Clément

INSERM, U-620, Detoxication and Tissue Repair Unit, University of Rennes I, 2 av. Léon Bernard, 35043 Rennes, France

Abstract

A variety of exogenous and endogenous agents, e.g., toxic compounds, drugs, pathogens, are responsible for acute and liver injuries which may lead to inflammation and fibrosis and impair hepatocyte functions. In inflamed and fibrotic livers, extracellular matrix remodelling is a complex mechanism of synthesis and degradation of matrix components, namely collagens, non-collagenous glycoproteins and proteoglycans. Fibrolysis is the result of the activation of proteases, among them matrix metalloproteinases, which cleave matrix components and release (poly)-peptide modules with specific biological activities. The dynamic turnover of extracellular matrix is regulated by cytokines and other soluble factors. Depicting these mechanisms opens the path to the identification of biomarkers and targeted drugs for the reversion of inflamed/fibrotic scar towards a normal architecture and the restoration of normal liver functions.

Introduction

The liver is constantly exposed to various endogenous and exogenous compounds and pathogens, which may induce acute and/or chronic injuries. These injuries lead to a complex process of tissue repair which consists in inflammation, extracellular matrix production and remodelling, and cell regeneration. In most cases, this process results in the restoration of the liver architecture without any loss of hepatic functions and obvious clinical signs. However, when liver injuries are repeated, tissue repair may become unsuccessful, a chronic inflammation process takes place and extracellular matrix components accumulate in excess with an increase of tissue remodelling which results in the formation of a fibrotic scar. Hepatic fibrosis is characterised by a collagen-rich matrix accumulation which spreads from injured areas towards the entire lobule and, subsequently between lobules thus resulting in the isolation of hepatocyte nodules. Such an accumulation of matrix components induces changes in the liver homeostasis, and reduces exchanges of soluble compounds between blood and hepatocytes. Cirrhosis is the end-stage of liver fibrosis.

Matrix Metalloproteinases in Tissue Remodelling and Inflammation,
edited by Vincent Lagente and Elisabeth Boichot
© 2008 Birkhäuser Verlag Basel/Switzerland

It is characterised by an abundant accumulation of matrix components, a disorganised liver architecture and vascularisation, and signs of cellular regeneration which occur within liver nodules. These nodules are the sites of preneoplastic lesions and dysplasia which are precursors of hepatocellular carcinomas whose 80% arises from a cirrhotic liver.

Fibrogenic cells

Most liver cells are involved in extracellular matrix remodelling, if not all [1, 2]. Among them, hepatic stellate cells (HSC) are central players: under most chronic injuries they undergo phenotypic changes towards a myofibroblastic phenotype, they proliferate and they produce extracellular matrix components, including interstitial (types I, III, VI) and basement membrane collagens (types IV and XVIII) non-collagenous glycoproteins (laminins) and proteoglycans. In addition, they produce various proteases at high levels, including matrix metalloproteases (MMPs) and other proteases [3]. Moreover, activated HSC synthesise endothelin 1 (ET-1) which regulates cell contractions, TGF-β1, a central cytokine in fibrogenesis, PDGF and MCP-1 which are involved in the recruitment of leucocytes and thus perpetuate inflammation. Stimulators of HSC activation include TGF-β1, PDGF, connective tissue growth factor (CCN2), whereas IL-10 and interferon-γ are anti-fibrogenic.

Not only HSC accumulate liver extracellular matrix components. Indeed, the heterogeneity of hepatic myofibroblasts suggests that they arise from various origins [4]. This was shown by analysing the expression of specific markers, e.g., glial fibrillary acid protein, fibroblast activation protein, α-smooth muscle actin, neural cell adhesion molecules, ICAM-1, PDGF receptor, nerve growth factor, neurotropin3, TRK3...which allowed the identification of subpopulation of myofibroblasts, probably originating from the peri-portal areas. Also, biliary epithelial cells and sinusoidal endothelial cells express basement membrane components, namely collagen IV, laminins and perlecan [5]; and hepatocytes produce fibronectins and collagen XVIII [6]. In chronic liver injuries, the expression levels of these matrix components are dramatically increased. Early works using immuno-electron microscopy suggested that hepatocytes might, under certain circumstances of injuries produce collagens I and IV [7], in addition to the recently discovered collagen XVIII [6]. These findings were recently confirmed by the demonstration of an epithelial-mesenchymal transition (EMT) in liver epithelial cells [8]. EMT involves gradual loss of epithelial phenotypic features, including cell-cell adhesion, tight junctions, E-cadherin, ZO-1 and cytokeratins. On the other hand the acquisition of a fibroblastic phenotype includes increased motility, expression of N-cadherin, vimentin, Snail, Slug, Twist and the production of collagen I, MMP-2 and MMP-9. EMT has been well documented in kidney, lung and mammary epithelial cells. In the liver, it has been shown that in response to TGF-β1, biliary epithelial cells [9] and hepatocytes [8] underwent EMT, gradually loosing epithelial phenotype towards fibroblastic morphology, including expression of collagen I and cell motility.

Matrix metalloproteinases in normal and fibrotic liver

Matrix metalloproteinases (MMPs) are endopeptidases containing an active site Zn^{2+} and divided into subfamilies [3, 10] which belong to the metzincin family, characterised by a three histidine zinc binding motif, together with ADAMs ('A Disintegrin And Metalloproteases'), astacins, serralysins, snapalysins and leishmanolysins. The MMP subfamily contains 25 members with structural identities, including conserved pro- and catalytic domains. In addition, they contain or not a furin-recognition motif. MMPs are either secreted or anchored within the cell surface.

Collagens I and III, the major collagen components of the hepatic fibrous scar are degraded by MMPs with interstitial collagenase activity. These MMPs, i.e., MMP-1, 2, 8 and 13 in humans and MMP-13 in rodents, cleave collagen at a single site [11]. This cleavage allows the collagen to partially unwind and renders it susceptible to degradation by more promiscuous MMPs and unspecific proteases, as well. Importantly, in the course of hepatic fibrogenesis, collagenolytic activity remains present but decrease when expressed relative to collagen content. It has been suggested that this may be related to the expression of specific inhibitors, namely the tissue inhibitors of matrix metalloproteinases, TIMPs 1 and 2 [10]. Freshly isolated HSC and early HSC primary cultures express MMP-1, MMP-13 and the uroplasminogen activator, uPA. When HSC become fully activated, this pattern of expression changes and cells express a combination of MMPs which can degrade normal liver matrix, while inhibiting the degradation of fibrillar collagens which accumulate in hepatic lesions. Thus, TIMP-1 expression dramatically increased in activated HSC, leading to a more global inhibition of degradation of fibrillar collagens by the interstitial collagenases, MMP-1 and MMP-13. *In vivo*, following cessation of liver injury, reversion of hepatic fibrosis may occur and this is related to changes in the pattern of MMP expression, while TIMP-1 is rapidly down-regulated [3, 12].

MMP-2 (gelatinase A) is involved in the early stages of extracellular matrix remodelling. MMP-2 cleaves collagen IV and other components of the basement membrane. In normal liver, MMP-2 mRNAs are expressed a low levels, being dramatically induced in fibrotic/cirrhotic livers and in both primary and secondary liver cancers, up to 300–400%. MMP-2 mRNAs co-localise to myofibroblastic-like cells and activated HSC [13, 14]. These cells also express MT1-MMP and TIMP-2 mRNAs, the main components of the MMP-2 activation process [15]. In addition, MMP-2 may act as an autocrine proliferative signal, thus increasing the number of both activated HSC and myofibroblasts, thereby contributing to a vicious circle towards fibrosis. When platted in primary culture HSC which become 'activated', express MMP-2 at high levels. Noteworthy is the upregulation of both synthesis and activation of MMP-2 by collagen I which has been demonstrated in HSC cultures, whereas collagen VI, laminin, and the reconstituted basement membrane matrigel were ineffective in inducing activation [16]. Indeed, cultured HSC secrete latent MMP-2, and MMP-2 activation occurs when they are associated with hepatocytes in co-culture, concomitantly with a collagen-rich matrix deposition [17].

ADAMs

Hepatic stellate cells are a main source of ADAMs, namely ADAM12 and 9 in fibrotic livers and in hepatocellular carcinomas [18, 19]. The ADAMs form a family of multi-domain glycoproteins highly homologous to the class III snake venom metalloprotease disintegrins [20, 21]. Up to 30 ADAMs have been identified with a broad tissue distribution, so far. They are involved in various biological processes, e.g., sperm-egg recognition and fusion, adipogenesis, neurogenesis and myocyte fusion. The common extracellular part of the proteins includes a regulatory prodomain and metalloprotease, disintegrin-like and cystein-rich domains. They are also characterised by an EGF-like domain, a transmembrane domain and a cytoplasmic tail.

ADAMs with potential sheddase activities have been identified, including ADAM12 which cleaves insulin like growth factor-binding protein-3 and -5 and the heparin-binding epidermal growth factor-like growth factor [22]. In addition, ADAM12 supports cell adhesion by interacting with syndecan and integrins, and has been involved in the differentiation process of myoblasts, osteoclasts and adipocytes.

TGF-β1 upregulates ADAM12 expression in HSC, probably through phosphatidylinositol 3-kinase (PI3K) and mitogen-activated protein kinase (MEK) inhibitor activities [23]. Importantly both ADAM12 and ADAM9 are upregulated in hepatocellular carcinomas and liver metastases from colonic carcinomas and correlates with an increase in MMP-2, suggesting an important role in tumour aggressiveness and progression [18].

Regulation of extracellular matrix remodelling

A number of cytokines are involved in the regulation of extracellular matrix remodelling. TGF-β1 is a central regulator of fibrogenesis and fibrolysis [24]. It is synthesised by platelets and Kupffer cells in the early stages of inflammation and then activated by MMPs or the tissue plasminogen activator. Thereafter, hepatic stellate cells produce TGF-β1 at high levels. TGF-β1 plays a key role in the activation process of HSC and upregulates matrix production, together with anti-proteases, e.g., α2-macroglobulin and the plasminogen activator inhibitor, PAI-1 [25]. Transcriptional activation of collagen I by TGF-β1 is mediated through SP1 and Smad3. Activated

HSC and myofibroblasts perpetuate the inflammation process by activating lymphocytes through chemokines, e.g., MCP-1 and RANTES [26].

Interestingly, antioxidant agents, e.g., vitamin E, may reduce fibrosis *in vivo* and *in vitro*. New therapeutic approaches, including specific antagonists targeted to cytokine and corresponding receptors are currently being explored, including PDGF, FGF and TGF-α. Targeting TGF-β1 should inhibit extracellular matrix production and enhance matrix degradation, particularly through TIMPs inhibition. Also, TNF-α antagonists may be anti-fibrotic as shown in arthritis and Crohn disease.

Eicosanoids constitute an extensive group of compounds which derive from C-20 polyunsaturated fatty acids, e.g., arachidonic acid, and include widely occurring tissue hormones such as prostaglandins, thromboxane, leukotrienes and other products of arachidonate oxidation [27]. Leukotrienes play a main role in liver tissue remodelling. Indeed, the liver not only represents the main organ for elimination of systemically produced cysteinyl leukotrienes, but it is also a production site of these pro-inflammatory mediators in the course of liver injury. Hepatic leukotriene generation can occur in at least three cell types, namely Kupffer cells, mast cells and under some conditions hepatocytes [28]. Inflammatory cells, such as neutrophils and monocytes, infiltrate injured liver and induce leukotriene production. Thus in endotoxin/D-galactosamine induced hepatoxicity, Kupffer cells are activated and followed by inflammatory infiltrations and hepatocyte necrosis which in turn may induce hepatitis. In this model, protection against fulminant hepatitis is achieved by inhibitors of leukotriene. Another example is liver damage which occurs following ingestion of the mushroom poison, α-amanitin combined with endotoxin. The resulting hepatotoxic effect can be prevented by dexamethasone which down-regulates leukotriene production in injured liver. In addition, hepatic hypoxia which may occur under conditions of reduced liver blood flow, sepsis, and shock, may potentate toxic effects of enhanced leukotriene concentrations in the liver. Interestingly, it has been shown that cysteinyl leukotrienes enhance the TNF-α-induced MMP-9 production *via* binding specific receptors in human monocytes/macrophages [29].

Proteolytic breakdown of extracellular matrix generates specific signals

In the pericellular area, the control of major cellular mechanisms depends on (poly-)peptide modules containing less than ten to several tens of amino acids within exposed domains of proteins, namely plasma proteins and extracellular matrix collagens, proteoglycans and non collagenous glycoproteins. Some of these modules, called 'matricryptins', are kept cryptic and thus latent within the parent molecule [30] (Tab. 1). They can be exposed, and thus activated by structural or conformational changes, and proteases which cleave the parent molecule, including MMPs. Such mechanisms are induced by the tissue microenvironment in diverse physiological or pathological situations [31]. For example, inflammation, fibrogenesis and

Table 1 - Bioactive modules encrypted within extracellular matrix components (modified from [30, 31])

Parent molecule	Module	Functions	Mechanisms of activation
Fibronectin	type 3 repeats	fibronectin assembly	mechanical forces
	RGD	cell adhesion	adsorption, mechanical forces, proteases
	120 kD and 40 kD binding domains	cell migration	proteases
	N-C-terminal fragments	inhibition of cell proliferation	proteases
Vitronectin	RGD	cell adhesion (αvβ3, αvβ5)	adsorption
Collagens	RGD	cell adhesion (αvβ3)	proteases: cathepsin G, collagenase
	Pro-Pro/Hyp-Gly	cell migration	proteases, denaturation
	C-terminal fragments	cell adhesion, angiogenesis	proteases
Collagen IV	α1 arm peptides	cell adhesion and migration	proteases
	NC1 domain Arresten, canstatin, tumstatin	cell adhesion (integrins)	proteases: MMP 9
Collagen XVIII	NC1-domain Endostatin	angiogenesis	proteases: elastase, cathepsins, MMPs
	FZC18	Wnt/β catenin signalling pathway	proteases
Decorin	Leucine-rich domain	EGF-receptor: cell signalling pathway	proteases
Laminins	γ2 chain fragment	cell adhesion and migration	proteases: MT1-MMP, MMP 2
	YIGSR, SIKVAV	cell adhesion	proteases
	EGF-L repeats	EGF-receptor, cell migration	proteases
Fibrinogen	γ chain, D domain	fibrin polymerisation	proteases
	B chain	vascular permeability	plasmin
Hyaluronic acid	3–16 fragments	angiogenesis	hyaluronidase
SPARC	KGHK	angiogenesis	proteases
	Helix loop A	collagen affinity	MMP
Osteopontin	N-terminal fragment	cell adhesion (α9β1)	proteases, thrombin
Thrombospondin	fragments	interactions with fibronectin cell adhesion	proteases

Table 1 (continued)

Parent molecule	Module	Functions	Mechanisms of activation
Tenascin-C	RGD	cell adhesion ($\alpha 8\beta 1$)	proteases
	EGF-like repeats	EGF receptor: cell migration/proliferation	proteases
Elastin	VGVAPG	cell migration	elastases
	(XGXXPG)	chemotaxis, cell proliferation	MMP 2, 9, 7, 12, elastase

tumorigenesis are accompanied by substantial extracellular matrix remodelling by MMPs, plasmin, cathepsins and other proteases, which in turn generate bioactive polypeptides. Several modules were recently identified which function as regulators of cell proliferation and apoptosis and angiogenesis, albeit differently from the parent molecules. Importantly, accessibility of these modules to plasma membrane receptors appears modulated through their interactions with other components, mainly glycosaminoglycans within the pericellular matrix.

Prototypical matricryptins include the epidermal growth factor-like (EGF-L) repeats of tenascin-C and laminins, the leucine-rich region of decorin, and the triple helical structure of collagens that bind DDR1 and DDR2 [31, 32]. These modules can activate the same receptor without sharing structural homology. Unlike the EGF-like repeats of laminin-5, the vast majority of the matricryptins are short peptide sequences, such as the IV-Hl domain from collagen IV [33]. Many utilise integrins as receptors in their proteolysed form, though with unique signalling properties. The majority of these modules bind to their receptors with low binding affinity. Nevertheless, cryptic modules are released as diffusible proteolytic fragments and their target cells are usually in close proximity which favours an increase in their effective concentration.

Collagen XVIII is a paradigm of multi-modular macromolecules which functions differently in the course of tissue remodelling. This particular collagen type contains several domains acting as bioactive polypeptides when they are cleaved from the parent molecule, namely endostatin a well-described potent inhibitor of tumour angiogenesis in mice, a thrombospondin-like module, a module with unknown function, and a *frizzled*-like module [34–36]. Collagen XVIII is expressed as three distinct variants by two separate promoters and alternative splicing of one of the transcripts [37]. Promoter #1 generates variant #1 which is a ubiquitous structural basement membrane component [38, 39]. Alternative splicing of transcripts from promoter #2 generates variants #2 and #3, which are secreted under the control of both liver-specific and ubiquitous transcription factors. The variant #3 carries FZC18 domain which is a 235-aa stretch with 10 conserved cysteines, bearing sequence and structural identities with the cysteine-rich domain of the extracellular domain of the *frizzled* receptors and the secreted *frizzled*-related proteins (SFRPs). Analysis of the function of the variant 3 of collagen XVIII revealed interesting features [40]. This variant is expressed at low levels in normal adult human tissues, being upregulated in fibrogenesis and in small well-differentiated liver tumours, but decreased in advanced human liver cancers. It is proteolytically processed into a cell surface FZC18-containing precursor that binds Wnt3a *in vitro* through FZC18 and suppresses Wnt3a-induced stabilization of β-catenin. Ectopic expression of either FZC18 or its precursor inhibits Wnt/β-catenin signalling, thus down regulating major cell cycle checkpoint gatekeepers cyclin D1 and *c-myc* and reducing tumour cell growth. By contrast, the full-length variant is unable to inhibit Wnt signalling, thus indicating that the signal is encrypted and released by enzymatic processing. Thus, collagen XVIII may have a dual effect: (i) as a heparan-sulfate proteoglycan, it could enhance Wnt signalling by raising local concentrations of Wnt at the cell surface, possibly promoting Wnt interaction with the FZ-LRP receptor complex; (ii) alternatively, during extracellular matrix remodelling, i.e., when proteases become active, FZC18 is released from its parent molecule, and it antagonises the interactions of Wnt with the FZ-LRP receptor, which in turn inhibits Wnt/β-catenin signalling.

Conclusion

Recent breakthroughs in liver biology and genomics gave new insights into our understanding of the molecular mechanisms which underlie the regulation of extracellular matrix remodelling in normal, inflamed and fibrotic livers. These approaches should help to design efficient tools for the detection and outcome of liver diseases, as well as innovative drugs targeted to the microenvironment [41] in order to stop, and possibly reverse, liver damages. This requires the development of accurate biomarkers capable of improving the monitoring, diagnosis and prognosis of chronic liver diseases. In addition, these biomarkers are mandatory for investigating the genetic and genomic alterations occurring in injured liver for the design of candidate-drugs, as well as for evaluating both their efficacy and safety in preclinical and clinical studies.

Acknowledgments

Personal works were supported by INSERM, University of Rennes I, ARC, National Institute of Cancer (INCa) and MAE (STAR programme).

References

1   Guo J, Friedman SL (2007) Hepatic fibrogenesis. *Semin Liver Dis* 27: 413–426
2   Clement B, Levavasseur F, Loreal O, Lietard J, L'Helgoualc'h A, Guillouzo A (1993). Role of hepatocytes and Ito cells in the synthesis and deposition of extracellular matrix. In: C Surrenti, A Casini, S Milani, M Pinzani (eds): *Fat-storing cells and liver fibrosis.* Kluwer Academic Publishers, Dordrecht, 13–22
3   Benyon RC, Arthur MJ (2001) Extracellular matrix degradation and the role of hepatic stellate cells. *Semin Liver Dis* 21: 373–384
4   Magness ST, Bataller R, Yang L, Brenner DA (2004) A dual reporter gene transgenic mouse demonstrates heterogeneity in hepatic fibrogenic cell populations. *Hepatology* 40: 1151–1159
5   Rescan PY, Loreal O, Hassell JR, Yamada Y, Guillouzo A, Clement B (1993) Distribution and origin of the basement membrane comportent perlecan in rat liver and primary hepatocyte culture. *Am J Pathol* 142: 199–209
6   Musso O, Rehn M, Saarela J, Theret N, Lietard J, Hintikka E, Lotrian D, Campion JP, Pihlajaniemi T, Clement B (1998) Collagen XVIII is localized in sinusoids and basement membranes and expressed by hepatocytes and activated stellate cells in fibrotic human liver. *Hepatology* 28: 98–107
7   Clement B, Grimaud JA, Campion JP, Deugnier Y, Guillouzo A (1986) Cell types involved in the production of collagen and fibronectin in normal and fibrotic human liver. *Hepatology* 6: 225–234
8   Zeisberg M, Yang C, Martino M, Duncan MB, Rieder F, Tanjore H, Kalluri R (2007) Fibroblasts derive from hepatocytes in liver fibrosis *via* epithelial to mesenchymal transition. *J Biol Chem* 282: 23337–23347
9   Roberston H, Kirby JA, Yip WW, Jones DE, Burt AD (2007) Biliary epithelial-mesenchymal transition in posttransplantation recurrence of primary biliary cirrhosis. *Hepatology* 45: 977–981
10   Iredale JP (1997) Tissue inhibitors of metalloproteinases in liver fibrosis. *Int J Biochem Cell Biol* 29: 43–54
11   Aimes RT, Quigley JP (1995) Matrix metalloproteinase-2 is an interstitial collagenase. Inhibitor-free enzyme catalyzes the cleavage of collagen fibrils and soluble native type I collagen generating the specific 3/4- and 1/4-length fragments. *J Biol Chem* 270: 5872–5876
12   Elsharkawy AM, Oakley F, Mann DA (2005) The role and regulation of hepatic stellate cell apoptosis in reversal of liver fibrosis. *Apoptosis10*: 927–939
13   Musso O, Theret N, Campion JP, Turlin B, Milani S, Grappone C, Clement B (1997) *In situ* detection of matrix metalloproteinase-2 (MMP2) and the metalloproteinase inhibitor TIMP2 transcripts in primary hepatocellular carcinomas and in liver metastasis. *J Hepatol* 26: 593–605
14   Milani S, Herbst H, Schuppan D, Grappone C, Pellegrini G, Pinzani M, Casini A, Cal-

abro A, Ciancio G, Stefanini F et al (1994) Differential expression of matrix-metallopro-
teinase-1 and -2 genes in normal and fibrotic human liver. *Am J Pathol* 144: 528–537

15 Theret N, Musso O, L'Helgoualc'h A, Campion JP, Clement B (1998) Differential
expression and origin of membrane-type 1 and 2 matrix metalloproteinases (MT-
MMPs) in association with MMP2 activation in injured human livers. *Am J Pathol* 153:
945–954

16 Theret N, Lehti K, Musso O, Clement B (1999) MMP2 activation by collagen I and
concanavalin A in cultured human hepatic stellate cells. *Hepatology* 30: 462–468

17 Loreal O, Levavasseur F, Fromaget C, Gros D, Guillouzo A, Clement B (1993) Coopera-
tion of Ito cells and hepatocytes in the deposition of an extracellular matrix *in vitro*. *Am
J Pathol* 143: 538–544

18 Le Pabic H, Bonnier D, Wewer UM, Coutand A, Musso O, Baffet G, Clement B, Theret
N (2003) ADAM12 in human liver cancers: TGF-beta-regulated expression in stellate
cells is associated with matrix remodeling. *Hepatology* 37: 1056–1066

19 Kesteloot F, Desmoulière A, Leclercq I, Thiry M, Arrese JE, Prockop DJ, Lapière CM,
Nusgens BV, Colige A (2007) ADAM metallopeptidase with thrombospondin type 1
motif 2 inactivation reduces the extent and stability of carbon tetrachloride-induced
hepatic fibrosis in mice. *Hepatology* 46: 1620–1631

20 Wolfsberg TG, Primakoff P, Mykes DG, White JM (1995) ADAM, a novel family of
membrane proteins containing A Disintegrin And Metalloprotease domain: multifunc-
tionnl functions in cell–cell and cell–matrix interactions. *J Cell Biol* 131: 275–278

21 Flannery CR (2006) MMPs and ADAMTSs: functional studies. *Front Biosci* 11: 544–
569

22 Mori S, Tanaka M, Nanba D, Nishiwaki E, Ishiguro H, Higashiyama S, Matsuura N
(2003) PACSIN3 binds ADAM12/meltrin alpha and up-regulates ectodomain shedding of heparin-binding epidermal growth factor-like growth factor. *J Biol Chem* 278:
46029–46034

23 Le Pabic H, L'Helgoualc'h, Coutand A, Wewer UM, Baffet G, Clement B, Theret N
(2005) Involvement of the serine/threonine p70S6 kinase in TGF-β1-induced ADAM12
expression in activated human hepatic stellate cells. *J Hepatol* 43: 1038–1044

24 Gressner AM (2002) Roles of TGF-beta in hepatic fibrosis. *Front Biosci* 7: 793–807

25 Knittel T, Fellmer P, Ramadori G (1996) Gene expression and regulation of plasminogen
activator type 1 in hepatic stellate cells of rat liver. *Gastroenterology* 111: 745–754

26 Vinas O, Bataller R, Sancho-Bru P, Ginès P, Berenguer C, Enrich C, Nicolás JM, Ercilla
G, Gallart T, Vives J et al (2003) Human hepatic stellate cells show features of antigen-
presenting cells and stimulate lymphocyte proliferation. *Hepatology* 38: 919–929

27 Corey EJ, Niwa N, Falck JR, Mioskowski C, Arai Y, Marfat A (1980) Recent studies on
the chemical synthesis of eicosanoids. *Adv Prostaglandin Thromboxane Res* 6: 19–25

28 Decker K (1985) Eicosanoids, signal molecules of liver cells. *Sem Liver Dis* 5: 175–
190

29 Ichiyama T, Kajimoto M, Hasegawa M, Hashimoto K, Matsubara T, Furukawa S

(2007) Cysteinyl leukotrienes enhance tumour necrosis factor-alpha-induced matrix metalloproteinase-9 in human monocytes/macrophages. *Clin Exp Allergy* 37: 608–614

30   Schenk S, Quaranta V (2003) Tales from the cryptic sites of the extracellular matrix. *Trends Cell Biol* 13: 366–375

31   Davis GE, Bayless KJ, Davis MJ, Meininger GA (2000) Regulation of tissue injury responses by the exposure of matricryptic sites within extracellular matrix molecules. *Am J Pathol* 156: 1489–1498

32   Vogel W, Gish GD, Alves F, Pawson T (1997) The discoidin domain receptor domain receptor tyrosine kinases are activated by collagen. *Mol Cell* 1: 13–23

33   Chelberg MK, McCarthy JB, Skubitz AP, Furcht LT, Tsilibary EC (1990) Characterization of a synthetic peptide from type IV collagen that promotes melanoma cell adhesion, spreading, and motility. *J Cell Biol* 111: 261–270

34   Rehn M, Pihlajaniemi T (1994) Alpha 1(XVIII), a collagen chain with frequent interruptions in the collagenous sequence, a distinct tissue distribution and homology with type XV collagen (1994) *Proc Natl Acad Sci USA* 91: 4234–4238

35   Muragaki Y, Timmons S, Griffith CM, Oh SP, Fadel B, Quertermous T, Olsen BR (1995) Mouse Col18a1 is expressed in a tissue-specific manner as three alternative variants and is localized in basement membrane zones. *Proc Natl Acad Sci USA* 92: 8763–8767

36   Saarela J, Rehn M, Oikarinen A, Autio-Harmainen H, Pihlajaniemi T (1998) The short and long forms of type XVIII collagen show clear tissue specificities in their expression and location in basement membrane zones in humans. *Am J Pathol* 153: 611–626

37   Lietard J, Theret N, Rehn M, Musso O, L'Helgoualc'h A, Dargere D, Pihlajaniemi T, Clement B (2000) The promoter of the long variant of collagen XVIII, the precursor of endostatin, contains liver-specific regulatory elements. *Hepatology* 32: 1377–1385

38   Musso O, Theret N, Heljasvaara R, Rehn M, Turlin B, Campion JP, Pihlajaniemi T, Clement B (2001) Tumor hepatocytes and basement membrane-producing cells specifically express two different forms of the endostatin precursor, collagen XVIII, in human liver cancers. *Hepatology* 33: 868–876

39   Musso O, Rehn M, Theret N, Turlin B, Bioulac-Sage P, Lotrian D, Campion JP, Pihlajaniemi T, Clement B (2001) Tumor progression is associated with a significant decrease in the expression of the endostatin precursor collagen XVIII in human hepatocellular carcinomas. *Cancer Res* 61: 45–49

40   Quelard D, Lavergne E, Hendaoui I, Elamaa H, Tiirola U, Heljasvaara R, Pihlajaniemi T, Clement B, Musso O (2008) A cryptic Frizzled module in cell-surface collagen 18 inhibits Wnt/β-catenin signaling. *PLoS One* 3: e1878

41   Kong HJ, Mooney DJ (2007) Micoenvironmental regulation of biomacromolecular therapies. *Nat Rev Drug Discov* 6: 455–463

Index

The PIR-Series
Progress in Inflammation Research

Homepage: http://www.birkhauser.ch

Up-to-date information on the latest developments in the pathology, mechanisms and therapy of inflammatory disease are provided in this monograph series. Areas covered include vascular responses, skin inflammation, pain, neuroinflammation, arthritis cartilage and bone, airways inflammation and asthma, allergy, cytokines and inflammatory mediators, cell signalling, and recent advances in drug therapy. Each volume is edited by acknowledged experts providing succinct overviews on specific topics intended to inform and explain. The series is of interest to academic and industrial biomedical researchers, drug development personnel and rheumatologists, allergists, pathologists, dermatologists and other clinicians requiring regular scientific updates.

Available volumes:
T Cells in Arthritis, P. Miossec, W. van den Berg, G. Firestein (Editors), 1998
Medicinal Fatty Acids, J. Kremer (Editor), 1998
Cytokines in Severe Sepsis and Septic Shock, H. Redl, G. Schlag (Editors), 1999
Cytokines and Pain, L. Watkins, S. Maier (Editors), 1999
Pain and Neurogenic Inflammation, S.D. Brain, P. Moore (Editors), 1999
Apoptosis and Inflammation, J.D. Winkler (Editor), 1999
Novel Inhibitors of Leukotrienes, G. Folco, B. Samuelsson, R.C. Murphy (Editors), 1999
Metalloproteinases as Targets for Anti-Inflammatory Drugs,
    K.M.K. Bottomley, D. Bradshaw, J.S. Nixon (Editors), 1999
Gene Therapy in Inflammatory Diseases, C.H. Evans, P. Robbins (Editors), 2000
Cellular Mechanisms in Airways Inflammation, C. Page, K. Banner, D. Spina (Editors), 2000
Inflammatory and Infectious Basis of Atherosclerosis, J.L. Mehta (Editor), 2001
Neuroinflammatory Mechanisms in Alzheimer's Disease. Basic and Clinical Research,
    J. Rogers (Editor), 2001
Inflammation and Stroke, G.Z. Feuerstein (Editor), 2001
NMDA Antagonists as Potential Analgesic Drugs,
    D.J.S. Sirinathsinghji, R.G. Hill (Editors), 2002
Mechanisms and Mediators of Neuropathic pain, A.B. Malmberg, S.R. Chaplan (Editors), 2002
Bone Morphogenetic Proteins. From Laboratory to Clinical Practice,
    S. Vukicevic, K.T. Sampath (Editors), 2002
The Hereditary Basis of Allergic Diseases, J. Holloway, S. Holgate (Editors), 2002
Inflammation and Cardiac Diseases, G.Z. Feuerstein, P. Libby, D.L. Mann (Editors), 2003
Mind over Matter – Regulation of Peripheral Inflammation by the CNS,
    M. Schäfer, C. Stein (Editors), 2003
Heat Shock Proteins and Inflammation, W. van Eden (Editor), 2003
Pharmacotherapy of Gastrointestinal Inflammation, A. Guglietta (Editor), 2004
Arachidonate Remodeling and Inflammation, A.N. Fonteh, R.L. Wykle (Editors), 2004
Recent Advances in Pathophysiology of COPD, P.J. Barnes, T.T. Hansel (Editors), 2004
Cytokines and Joint Injury, W.B. van den Berg, P. Miossec (Editors), 2004